TRIAL BY SOIL

TRIAL BY SOIL

PREHISTORIC

&

SUBMERGED PRE-CONTACT

GEOARCHAEOLOGY

ETHAN WILLIAM MELLON

Trial by Soil: Prehistoric & Submerged Pre-Contact Geoarchaeology

Published by Ethan William Mellon
First Edition: July 2026
Hardcover ISBN: 979-8-234-13304-5
Paperback ISBN: 979-8-234-13303-8

Printed in the United States of America

For my peers,

May you carry the lantern of illumination and bring light to those you touch.

CONTENTS

PREFACE

If I have seen further, it is by standing on the shoulders of giants.

- Isaac Newton

In the early 2000s, my mother and father used to take my younger brother and me to a beach near the Buffalo-Canadian border in New York when we visited my grandmother. It was a fake beach. One that became primarily formed during the glacial and interglacial periods of the last ice age as glaciers advanced and retreated, slowly carving out the depression basin that would eventually become Lake Erie. The Laurentide Ice Sheet, which covered much of the region, left meltwater from retreating glaciers filling in the Great Lakes over time. My parents used to tell me stories of when they sailed these lakes together.

When I was still a child, there was a day when we visited this pseudo-beach, my brother taking off to chase gulls with a few sticks and seaweed he touted to be a "bow and arrow." Like any savvy adventurer, I chased after him. Hopping along the litter and cigarette butts of Buffalo's finest "white sand beaches." We came to where the bluff met with the edge of the bay, where seemingly immense piles of driftwood lie along the shore. After clambering over the driftwood, along with washed-up trash and broken shells, I finally caught up to my brother, where he stopped along the shore. Out of breath and slouching over him, I traced his wide-eyed attention to the object poking out of the sand at his feet, bow and arrows sheathed knightly at his side. I reached down and picked up the first object I had encountered, clearly older than my own, if not my parents' existence, that was made by another human being left behind as an artifact from another time.

I turned it over in my hands. "VILLE," it read on the reverse side, half-broken like a Hellenistic bust and made of solid brick. For the first time, I felt a wave of solemn uncanniness rush over my shoulders from the sight of something so unequivocally human, lost to time. Echoes of the past, which were now present in my hands, created a sense of empathy and rueful yearning for some distant shared humanity. Was this from a former village...town...city that *they* had left behind here? Or did it wash in from a far-away land, only now peeking its hand out from history, reaching for someone or something to show it the way home.

I remember bringing it back to show my father, who at first took little interest in the object. Yet, it remained within our household for more than twenty years. As I write this today, I notice that it has become a doorstop for a closet inherited into my own home. However, as I've gotten older and the sciences have changed my perceptions of human capabilities for altruism, the object stands as another reminder: once you remove something from its context, the potential to learn about its geological and often historical context is lost forever. Bound by the will of the greedy. As anthropologist Loren Eiseley put it in his 1971 memoir, *The Night Country*, "A man who has once looked with the archaeological eye will never see quite normally. He will be wounded by what other men call trifles."

With many turning a blind eye to the atrocities being committed on world heritage sites, including the destruction of historically significant centers in the 21st century, these words should serve as a cautionary tale. Archaeologists spend their entire lives trying to understand our history and decipher the past, only for some fool to come and destroy it out of avarice. "Most voluntary crimes undoubtedly occur due to the ambition and greed of men" (Aristotle). Now more than ever, it is the responsibility of cultural heritage professionals to act and persevere in the preservation of our shared history. It is what brought me to the forefront of this science as a young student and compelled me to delve into the deepest recesses of human history to write this book.

In my first year as a student in archaeology, I came across a book documenting the 1929 zoological expedition of the first U.S. Forest Service Chief, Gifford Pinchot, *To the South Seas*. It details the efforts of Pinchot and his crew of scientists and multidisciplinary researchers on their expedition to the South

Pacific aboard the schooner "Mary Pinchot." The crew documented and collected specimens for the National Museum in Washington; however, the book also recounts Pinchot's observations and personal reflections on the flora, fauna, and people of each island they visited. In effect, the text follows the personal developments and reflections of Pinchot as he neared the end of his journey, written from his own perspective. His immersion in both the wildlife and cultures of the people he encountered in the South Pacific brought him to a deeper anthropological perspective by the end of his journey.

After I finished reading Pinchot's expedition in its entirety, I had a newfound sense of commitment to approach archaeology from a conservation perspective; however, I would have to tread carefully, or else risk falsely reifying processualist ideologies in an attempt to seek out any scientific techniques of those who came before me. Anthropological theory became a lantern in the darkness of modern ontological thought, which underscored much of my early career in archaeology. It was here that I not only sought to learn conservation methodologies of artifacts, but of the dynamics an environment plays on the long-term preservation of cultural remains. The influence of geological processes on the movement and behaviors of humans across cultural landscapes brought forth questions that dealt with deep time. Thus, exploring the way in which I could incorporate earth sciences and prehistoric archaeology in my studies revealed itself as a coherent and intellectually fulfilling direction. It soon rationalized itself before me as *geoarchaeology*. It was also here that I realized I could intertwine these branches of inquiry with beautiful symmetry in maritime archaeology. The draw of Pinchot's vision in environmental and heritage conservation, employed physically in the maritime field, was something I aspired to. "Conservation aims at the greatest good for the greatest number for the longest time" (G. Pinchot). As I continued throughout my early career in archaeological fieldwork, it slowly transformed into a necessity to understand human anatomy, for I learned the wilderness is a foreboding place.

Throughout my undergraduate program, I developed a keen sense of operational awareness after being injured or stranded more than once in the backcountry. In 2018, I became stranded in the Alaskan wilderness along the stretch of taiga between Denali and Nenana, without a compass and only broken down train tracks in the woods to follow my way out. I had embarked on this trek to Denali with my stepfather, who at the time was rather estranged from me, and it is where I first learned that you either find your way out or face the ultimate consequence. After passing through the taiga in the park on a day visit, we stopped for a break, waiting for the train back to Fairbanks. Naturally, falling asleep when you have a ticket to leave is a surefire way to get stuck in a national park at night. Later the same year, we ventured to the desert in New Mexico known as White Sands. I found that these xeric environments can not only be extremely hot during the day, but also excruciatingly cold at night when underprepared. In these scenarios, learning to care for others in the wilderness is just as important as watching out for yourself. Finally, we traveled across Italy and Ireland to dive the Ligurian Sea and traverse the geological marvel of Giant's Causeway, which further ignited my interest in submerged geo-archaeology. By the spring of 2019, I took my first trip without a helping hand to guide me. This time I was the light in the dark.

One of my dear friends in archaeology, and thankfully, along with his bilingual colleague, joined me for this specific trip to Havana, Cuba. We had set out to survey the historical Havana Channel near the site cleared by Captain Juan Bonilla before his fleet, including the merchant vessel *El Salvador*, was driven into the North Carolina coast during a 1750 hurricane. The wreckage, along with the cargo onboard, has yet to be discovered to this day. Some allege that the wrecked cargo was intercepted by merchant traders Owen and John Lloyd, potentially making this one of the most notable maritime interventions in history. It was this movement of Spanish silver on a routine trip home to Spain that inspired the legends of Long John Silver in Robert Louis Stevenson's *Treasure Island*. This independent investigation was the first time I had followed in the footsteps of Pinchot, undertaking the rigorous demands of maritime exploration to advance cultural conservation. This, to my knowledge, was the first attempted survey of the Havana Channel for any evidence of the departing *El Salvador* fleet since either the events of 1750 or the pioneering work of Gordon P.

Watts, who searched for the wreckage in the 1990s. As one of my earliest lessons in underwater survey, the project in Cuba was a stark reminder of human fragility. It prompted me to go to great lengths in order to understand human anatomy and gas exchange physics, an aspect crucial to diving.

What followed was a trip to Iceland a month later for wilderness experience, where I scaled part of the Sólheimajökull glacier and traversed Snæfellsnes Peninsula with two other close peers, one of whom was also in an anthropological discipline at the University of North Carolina at Charlotte like myself. It was somber following the tragic events at our university in April. As a child, I dedicated a significant amount of time competing in regional and state geology competitions, and I was finally able to witness a significantly active geological wonder of the world in all of its grandeur. A moment to step away from the chaos of human fragility that seemed to follow me everywhere I went. It was also the beginning of my solo mountaineering and wilderness medicine experience, in which my subsequent field school later that summer brought more of its importance to light. I would next be headed west of the Amazon rainforest and Andean mountain range to the Chongón-Colonche Cloud Forest of Ecuador. Here, I would study a prehistoric culture known as the Valdivia.

When June rolled around, I found myself deep in the rainforest learning practical field methodology of prehistoric South American archaeology. This would be the first time that feeling of uncanniness from Lake Erie washed over my shoulders again upon revealing the first incised pottery sherd in the soil. The *Erie Effect* was certainly an eerie feeling. Something between the uncanny valley and universally confronting the unknown, despite a vast time difference, which blends a sense of uneasiness and familiarity between us and the past. It violates our expectations as human beings. These perspective-taking moments of our shared humanity pull on the strings of empathy, and it is what ignited my curiosity into the odyssey of humankind.

Following my foundational field training, I dedicated myself to rigorous wilderness medicine courses. While working in the Ecuadorian rainforest, I fell into a bramble of barbed wire that lacerated my upper legs. This incident highlighted the critical necessity of medical preparation in remote environments, prompting my focused study on emergency medicine and its practical application in archaeological fieldwork. I eventually earned certifications as a wilderness first responder and wilderness survival instructor, later accepting a position in pharmaceutical medicine for several years while continuing my research. This background allowed me to combine clinical knowledge with wilderness medicine principles to further explore the intersection of safety and archaeology as a permanent career path. By late 2019, driven to establish a highly capable field unit, I moved with a level of vigor and swiftness that others seemed to fear in me. What followed at the end of that year was the creation of an independent archaeological research organization formed between a team of wilderness medicine professionals and field researchers.

Founded on the basis of bringing attention to the importance of medical knowledge and its practicality in archaeological fieldwork, in addition to documenting and preserving cultural hertiage sites acrross the Southeastern United States, an archaeological research organization was created in 2019 as a collaborative team of archaeologists and wilderness medicine professionals. While ambitious in scope, the core message was clear: to protect vulnerable heritage sites from unauthorized disturbance, practice archaeology safely, and most importantly, operate in an ethical manner that preserves critical context to produce credible and interpretable data.

For those who view antiquities through the lens of casual collecting or commercial trade, it is vital to understand what is lost in that process. Archaeologists dedicate their lives to analyzing the precise geological and stratigraphic contexts in which cultural remains are deposited. When an artifact is casually extracted from the sediment simply to be sold online, its historical context is permanently destroyed, erasing the critical data needed to interpret our shared human past. There is a better way to uncover the past that contributes in a meaningful way to our human and geological history. Becoming involved with your local State Historic Preservation Office is an invaluable place to start.

Ultimately, establishing this research initiative required an immense amount of dedication and fieldwork. Over the years, our team has carried out several systematic field investigations, utilizing probability-based GIS mapping, non-destructive remote sensing, and precision extraction techniques to document, recover, analyze, and stabilize artifacts, ensuring all findings meet the strict standards required by institutional and regulatory repositories. These specimens were processed through a dedicated conservation laboratory structured around methodologies gained through years in pharmaceutical medicine, my time at the Bioarchaeology Laboratory of UNC Charlotte, and the foundational material culture conservation manuals of Texas A&M University.

Today, we are a team of multidisciplinary researchers operating out of a central office and research laboratory located in the Southeastern United States. While focused in the Southeast, we are a dedicated force in the field of cultural resource preservation. Let this book be a testament to that. My time in the field has revolved primarily around prehistoric contexts and attempting to interpret the geological or biological conditions that affect them underwater. This has culminated in a deep understanding of the earth sciences, chemistry, biology, and physics, drawing me into the foundations of geoarchaeology. It gradually came to me as a responsibility rather than an initial desire to explore the subfield of geoarchaeology itself. Operating this organization required a duty to prioritize objective, rigorous science first and foremost. This required an extensive knowledge of geomorphology, stratigraphic sciences, paleoenvironments, paleontology, geophysics, and archaeometry, which transformed my understanding of maritime archaeology. I found myself at the edge of a new frontier marked by a profound responsibility, with years of experience in the field behind me guiding every bolt of thunder I called down. It became my personal *raison d'être* to facilitate a space where students could also learn this information and have a chance to conduct archaeological fieldwork when they did not, or could not, attend a traditional field school.

Following undergraduate studies, I took a position at an electrical engineering firm as an assembly technician alongside continuing work in pharmaceutical medicine simply to make a living. I maintained small-scale independent field work, but took time to engross myself in DC electrical theory. I built high- and low-horsepower DC drives for just over a year before fully committing to pharmaceutical medicine for a number of years following that. After learning basic Python coding in open source software, which allowed me to build predictive models in GIS, I brought my knowledge of electrical physics into the field by building a simple sonar setup engineered for stationary deployment. This allowed a certain previously unattained agility to our field investigations by combining geophysics and GIS software. I am now in the early stages of working on geophysical applications with theoretical and particle physics, which, if proven effective, may surpass the resolution limitations of ground-penetrating radar (GPR) in some geological contexts. As physicist N. David Mermin stated in a 1990 *Physics Today* article, "Particle physics has become the archaeology of physics."

What I have found throughout my time in the field is that students often feel overburdened with information when you start throwing words around regarding time periods. This is especially the case when jumping between large temporal periods of geologic and archaeological time, whether in conversation or instruction, which have completely separate contexts. Not to mention the daunting and sometimes overlooked differences between pedology (soil science) and stratigraphy. This book serves as a practical guide and reference manual on the history of geologic time, geologic time periods themselves, the prehistoric ages, archaeological periods, paleoanthropology, geophysics, archaeometry, and geoarchaeology. It is for those who seek the past in order to understand the future of humankind. May this knowledge serve you as well as it has served me.

TRIAL BY SOIL

PREHISTORIC & SUBMERGED PRE-CONTACT GEOARCHAEOLOGY

HISTORY OF GEOLOGIC TIME

There is no greater education than one that is self-driven.

- Neil deGrasse Tyson

OUR UNDERSTANDING OF EARTH'S GEOLOGICAL history is a culmination of relative and absolute dating methods that have been constantly refined over the past several centuries. Before these techniques it was nearly impossible to decipher Earth's age throughout much of human history. When geologists initially began attempting to estimate the age of the Earth, relative dating was the primary method, which relied on the chronological order of rock layers and the different species of fossils they contained. This could aid in determining which layers were older or younger than others, but could not assign approximate numerical ages to specimens until absolute dating methods became more developed.

Considered to have laid the early groundwork for the development of modern stratigraphy, Danish scientist Nicolaus Steno, born Niels Steensen (1638-1686), introduced many fundamental principles and concepts of geology. Originally studying medicine at the University of Copenhagen in the 1650s, he left Denmark in 1660 to continue his studies at the University of Leiden in the Netherlands. He briefly traveled during this time, studying in both Montpellier and Paris before moving in 1665 to Florence, Italy, where his research in anatomy facilitated his election to the scientific body Accademia del Cimento (University of California Museum of Paleontology 2025). In 1667, Steno's work shifted from medicine to a profound insight into paleontological and geological sciences when he published the roughly 40-page dissertation, *Elementorum Myologiae Specimen*. Here, he draws symmetries of teeth from a shark captured from the town of Livorno, *Carcharodon rondeletii*, to the similarities of what was generally understood as fossils during this time, *glossopetrae*. Throughout the early 17th century, a lack of general geological publications led to the adoption of the term *glossopetrae*, also known as tongue stones; it referred to the idea of fossils naturally growing in rocks. This came with a general superstition that they contained preternatural properties.

Steno argued in his publication that fossils had actually come from formerly living organisms that had been buried by sediment deposits and geological processes over long periods of time. In justification, he used the Corpuscular Theory of Matter, now replaced by fundamental particle physics, to explain why fossils could change their chemical compositions without altering their physical characteristics (Steensen 1667). With new ideas about the processes of geological formations and their relationship to fossils, in 1669, Steno published *Preliminary Discourse to a Dissertation on a Solid Body Naturally Contained*

Within a Solid, in which he put forth many of the early principles of stratigraphy derived from his interpretations. In this publication, he defines rock strata by explaining the formation processes of sedimentary layers deposited horizontally as a cementation of particles settling from fluids over long periods of time. Recognized as Steno's law of original horizontality, it also put forth that any changes in a layer's position deposited this way were caused by future disturbances. Steno made significant contributions to crystallography in this dissertation as well, with the law of constant interfacial angles (Rafferty et al. 2013). This describes the corresponding faces of homogeneous minerals as always exhibiting the same angles regardless of size and appearance.

His most famous contribution to geology in this publication was his law of superposition, which made an argument to explain the natural stratigraphic order of Earth for the first time. Steno posited that upper layers of stratigraphy take the shape of lower layers, representing their temporal distinctions as upper layers being younger and lower layers being older deposits unless disturbances change this order. This was correlated from his observation that fossils left their imprints on softer sedimentary deposits, suggesting the fossil itself to be older than the surrounding rock. Another key principle put forth in Steno's publication was lateral continuity, in which he argued that sedimentary layers are continuous and widespread unless disturbed by a barrier or erosion process, indicating they were once connected.

Furthermore, he developed the idea of cross-cutting relationships in his paper, which stated that geological disturbances must be younger than features or formations affected by the disruption (Steensen 1669). Steno's brief but important contributions to geology have had a lasting impact in the earth sciences, providing a foundation upon which the understanding of Earth's stratigraphy could be built.

Roughly a century later, Giovanni Arduino (1714-1795), considered the father of Italian geology, would go on to develop Steno's early ideas of stratigraphy further. Arduino was an experienced field geologist and renowned miner in Vicenza by the time he published his first theories on stratigraphy in 1760, *Two Letters on Various Natural Observations*, to an Italian journal. Here Arduino outlined his interpretations of rock layers into Primary (Paleozoic; containing schist and crystalline rock), Secondary (Mesozoic; comprised of limestone and fossil), Tertiary (Cenozoic; primarily clay, volcanic depositions, and sandstone containing fossils), and a provisional fourth layer representing the softer alluvial deposits of the more recent epoch (Vaccari 2006). Originally sent as a series of letters to naturalist Antonio Vallisneri Jr. in 1759, these observations stemmed from his study of rock stratigraphy in the Algo valley in 1758 (Ell 2011, 2012). Arduino's insights into the formation processes of the Italian Alps laid much of the groundwork for further iterations of chronostratigraphy. This included later defining his posited notion of a fourth youngest layer of alluvium and estuarine deposits by H. Reboul in 1833, to which Reboul credits French geologist Jules Desnoyers (1800-1887) "as being first to use the term *Quaternaire*, which he used unequivocally to distinguish the deposits from those of the Tertiary" (Gibbard 2019, 11-19). This term was identified with deposits representing the Quaternary in Earth's geochronology.

It was not until 1788 that more theoretical foundations for modern geology were put forth by Scottish geologist James Hutton (1726-1797), in his publication *Theory of the Earth*. Hutton asserted the principle of uniformitarianism, stating that all geological processes must be gradual, have been consistent over time, and that changes to the Earth's stratigraphy resulted from continuous and uniform geological processes (Bushman 1983). This challenged earlier conjectures of catastrophism in Swiss naturalist Charles Bonnet's (1720-1793) *Philosophical Revival*, published in 1769, which put forth the idea that throughout different periods of Earth's history, all living things were subjected to a catastrophic extinction event, replacing all lifeforms with completely different populations. Hutton's theory, however, held "that the geologic forces and processes that shaped the Earth in the geologic past were referable to the same ones still in operation on the Earth's surface and could be observed directly" (Koutsoukos 2005, 7). The fundamental concepts from his theories facilitated the general acceptance of distinctive geological time sequences. Establishing uniformitarianism as a core principle of geology, British geologist John Playfair

(1748-1819) published a simplified version of Hutton's work, *Illustrations of the Huttonian Theory of the Earth*, for the general public in 1802 (Playfair 1802).

The realizations Hutton had observed in Earth's stratigraphy through uniformitarianism were further developed in the early 19th century by English geologist William Smith (1769-1839) between 1816 and 1819. He studied various sedimentary rocks and their distinctive fossilized remains, to which "Smith noted that the sequence of fossils in any given stratigraphic record follows a specific order, as a result of evolution; and that the same sequence can be found in isolated strata elsewhere, and thus correlated between them" (Koutsoukos 2005, 8). This observation led to the principle of faunal succession in *Strata Identified by Organized Fossils*, which challenged many naturalistic and religious doctrines associated with catastrophism during the late 18th through mid-19th centuries (Smith 1816). Smith's publication only solidified his earlier work, where he used the principle of faunal succession to publish the first lithostratigraphic map of southern England, Wales, and parts of Scotland in 1815. This was considered a revolutionary advancement in geology at the time, contributing significantly to many of the derived methods in chronostratigraphy today.

Around this same time, English geologist Etheldred Benett (1776-1845) and her contemporary, English paleontologist Mary Anning (1799-1847), were making significant contributions to techniques in biostratigraphy using evidence from fossils. Benett collected vast amounts of geological and paleontological specimens, with the earliest in her records from 1809-1810. In 1815, she compiled *The Measure of the Different Beds of Stone in Chicksgrove Quarry in the Parish of Tisbury, Wiltshire,* which was the first stratigraphic measured section of limestone from Upper Chicksgrove Quarry in Wiltshire, England (Burek 2001). This was illustrated and sent to The Geological Society; however, it was later published in 1816 under J. Sowerby without accreditation of Benett's work. Anning is known for her contributions to paleontology as a science through her discovery of the first complete skeletons of both an ichthyosaur, around 1836 (Oxford University Museum of Natural History 2025), and a plesiosaur discovered in the English coastal town of Lyme Regis in 1823. Her work also involved the discovery and study of pterosaur skeletons, further contributing to biostratigraphy of the Mesozoic Era.

Considered to be the father of geoarchaeology for his role in exploring the methods by which geology could aid in the understanding of past human activity, Scottish geologist Sir Charles Lyell (1797-1875) further developed Hutton's idea of uniformitarianism. In 1830, he published his widely influential book, *Principles of Geology*, in which he argued that past geological events could be explained by the same consequential processes we observe today, in an attempt to refine Hutton's theory of uniformitarianism. By his third edition of Principles in 1833,

Lyell proposed dividing the Tertiary Period into four subdivisions, the Eocene, Miocene, Pliocene, and Recent, based on fossil compositions of stratigraphy he had studied. He would also term the Pleistocene as the fourth subdivision in 1839 to distinguish the more recent fossil layer from the Pliocene. Just before this, in 1838, Swiss geologist Amanz Gressley coined the term "facies" to separate clearly and distinctly the physical and biological aspects of rocks from their age, which fundamentally contributed to our modern understandings of paleogeographic reconstruction (Cross and Homewood 1997). The terms Paleozoic, Mesozoic, and Cenozoic were later derived from Lyell's definitions by English geologist John Phillips, to define geological eras when he published the first Geologic Time Scale in 1841.

Lyell's work became an inspiration for Charles Darwin (1809-1882), with whom Lyell became acquainted. It was not until 1859 that Darwin published *On the Origin of Species*, after encouragement to publish his theories from Lyell (Edinburgh Geological Society 2020). With new interest in human evolutionary ancestry from Darwin, along with the immense geologic time posited by his earlier work in Principles, Lyell culminated these insights into his 1863 landmark book, *The Geological Evidences of the Antiquity of Man*. This groundbreaking publication played a pivotal role in establishing evidence for past humans and the importance of geological evidence in understanding prehistory. Lyell draws on symmetries from Darwin in his work stating, “By the multiplying of slight modifications in the course of thousands of generations, and by the handing down of the newly-acquired peculiarities by inheritance, a greater and greater divergence from the original standard is supposed to be effected, until what may be called a new species, or, in a greater lapse of time, a new genus, will be the result” (Lyell 1863, 411). Since Lyell’s publication, the Earth sciences and prehistoric archaeology have gradually come to a symmetry by uprooting the rather dogmatic conjectured approaches to chronostratigraphy of the 17th and 18th centuries associated with catastrophism.

At the turn of the 20th century, significant advancements in nuclear physics were evolving, which widely contributed to the development of absolute dating methods such as radiometric dating by American chemist and physicist Bertram Boltwood (1870-1927). Considered the father of nuclear physics, New Zealand physicist Ernest Rutherford (1871-1937) laid crucial groundwork for understanding radioactivity. After attending a lecture on radioelement dating given by Rutherford at Yale in 1904, Boltwood set out to determine the non-radioactive end product of uranium decay. By detecting trace radioactive elements and their corresponding decayed end-product, a geological specimen's age could be calculated based on the half-life of the element present. While analyzing ore and mineral samples in his private laboratory, he found that all samples containing uranium also contained traces of lead, concluding that lead must be the final product of the uranium series with radium as an intermediate step (Boltwood 1907, 77-88). From there, Boltwood was able to demonstrate how this technique could be used to determine the age of rocks and minerals to provide a more precise range of geological time periods. While his initial estimated Earth age calculations in 1907 of 2.2 billion years were off, he is often attributed as pioneering the use of radiometric dating based on uranium-lead decay, which was later improved on by British geologist Arthur Holmes (1890-1965). It was only 4 years later, on January 6, 1912, when German geologist Alfred Wegener (1880-1930) proposed his theory of continental drift to the German Geological Society. His famous hypothesis proposed that the continents had once been conjoined into the supercontinent Pangea and slowly drifted apart. Wegener's work is often credited as critical to our development in understanding plate tectonics.

Throughout the 20th century, the coordination of chronostratigraphic methods and prehistoric archaeology first gained traction through publications like Karl W. Butzer's "Geo-archaeological interpretation of Acheulian calc-pan sites at Doornlaagte and Rooidam (Kimberley, South Africa)," published in 1974 in the *Journal of Archaeological Science* (Butzer 1974). This would be the first specific mention of the term geoarchaeology after geological approaches had been implemented into the study of historic-period archaeological sites during the earlier second half of the century. As geoarchaeology began transforming both laboratory and archaeological fieldwork throughout the late 1970s and 1980s, "its status as a scientific discipline became concrete with the creation of the journal *Geoarchaeology* in 1986" (Buezen-Waller, Stock, and Kondo 2018). It is now widely implemented in archaeological inquiry through both innovative technology in geophysics and various scientific specializations that further geochemical analytic methods. Moreover, the applications of prehistoric and submerged pre-contact geoarchaeology are now fundamentally shaping our contemporary understandings of paleoenvironments and their effects on past human settlement patterns through the lens of these specializations and their associated technologies. It is with sincere encouragement that the knowledge and concepts of geological time, prehistory, archaeological periods, pedology, stratigraphy, and the methods in which we cross-examine with geoarchaeology are cultivated to a second nature by the practicing archaeologist.

THE GEOLOGIC TIME SCALE

Any fool can know. The point is to understand.

- Albert Einstein

THE INTERNATIONAL UNION OF GEOLOGICAL Sciences (IUGS) was founded in 1961 to coordinate research between the International Geological Congress (IGC), which was held every 4 years. Prior to the IUGS, the primary mechanism for discussion around geological research and collaboration internationally was the IGC. In 1974, the International Commission on Stratigraphy was founded, which is the oldest and largest scientific body in the IUGS.

After centuries of debate and scientific research, the Geologic Time Scale (GTS) as we recognize it today is organized into five distinct primary subgroups: Eons, Eras, Periods, Epochs, and Ages. These groups are characterized by the International Chronostratigraphic Chart, which was developed by the International Commission on Stratigraphy, and it serves as the standard for geochronology and chronostratigraphy.

Eons

Eons (chronostratigraphic eonothems) are considered the largest slice of geologic time and range from half a billion to nearly two billion years long. They are defined by significant transformations in Earth's geological history and biological evolution. The four recognized Eons in Earth's history are the Hadean, Archean, Proterozoic, and Phanerozoic.

Eras

Eras (chronostratigraphic erathem) are subdivisions of eons in geologic time. They typically span tens to hundreds of millions of years and are the second-largest unit in the geologic time scale. Eras are significant time periods in Earth's history, each distinguished by changes in the Earth's environment, evolution and diversity, and geological events. This includes the formations of mountains, fluctuations in sea level, and mass extinction events which distinguish eras from one another.

Periods

Periods are subdivisions of eras that represent smaller geologic time distinctions, typically hundreds of millions of years long. They are defined based on climate events, changes in stratigraphic fossil composition, and changes in Earth's geography. Specific stratigraphic rock layers and systems are associated with particular periods, where changes in dominant life forms reflected in the fossil record serve as a geological marker for periods.

Epochs

Epochs (chronostratigraphic series) are subdivisions of periods and represent less significant changes in stratigraphy than periods, ranging from hundreds of thousands to millions of years. The Neogene and Quaternary periods both contain two further subdivisions of epochs. Changes in rock layering associated with major environmental and climate shifts, such as glacial cycles, can mark epoch boundaries.

Ages

These are the smallest subdivisions of geologic time lasting anywhere from thousands to millions of years. They are defined by drastic shifts in the Earth's climate, such as a glacial or interglacial period within a glacial cycle, which is often reflected in the fossil record. This is correlated across different regions based on the principle of faunal succession. Ages capture more specific details within an epoch, with our current being the Meghalayan Age of the Holocene Epoch.

Hadean Eon

4.567 to 4.031 billion years ago

The Hadean is considered Earth's earliest eon, a time of constant bombardment from comets and other celestial bodies that deposited a wide range of elements and minerals, which have since subducted under tectonic movement. Earth's core and crust were forming and stabilizing during this time, which facilitated intense volcanic activity and a partially molten surface. This period of tectonic instability and volcanism on Earth consequently earned the Hadean Eon its name after Hades, the god of the underworld in Greek mythology.

The early atmosphere was primarily composed of carbon dioxide, nitrogen, and water vapor, among other gases, including ammonia and methane. This eon does not have formally recognized eras or periods by the International Commission on Stratigraphy due to the lack of available rock and fossil records from this time period. However, the Global Standard Stratigraphic Age (GSSA), used to define the base of a chronostratigraphic unit, is 4567.30±0.16 Ma for the lower boundary of the Hadean according to the International Commission on Stratigraphy. This is based on some of the oldest solid condensates known to have formed in the solar system, found within both the Allende (located in Mexico) and Efremovka (located in

Kazakhstan) meteorites, classified as calcium-aluminum-rich inclusions (CAIs) (Halla et al. 2024). Earth was still in the early stages of forming, undergoing intense heating and high atmospheric pressure, which made it difficult for stable rock formations to solidify and persist.

The thin upper layer of Earth during this time, along with the rampant volcanic activity, contributed to constant crustal recycling, which melted and reprocessed Hadean rock layers. This continuous destruction and tectonic plate recycling resulted in the lack of available Hadean rock in the geological record. However, detrital zircon crystal grains found at the Jack Hills conglomerate in Western Australia were dated using the uranium-lead (U-Pb) method, revealing their age of approximately 4.4 billion years (Wilde et al. 2001). The radioactive decay ratio of uranium to lead within the zircon crystals was analyzed using a sensitive high-resolution ion microprobe (SHRIMP II) to determine their age. This geochemical analysis instrument sputters ions from a material's surface and analyzes them based on their mass-to-charge ratio using a high-resolution mass spectrometer.

The age identified was older than any previously known zircon material that had been dated by about 130 million years. The high oxygen isotope ratios ($\delta^{18}O$) and rare earth element composition found in these zircon samples also indicated that the crystals formed from granitic magmas, which interacted with reworked continental material on the surface altered by water (Mojzsis et al. 2001). These findings implied that Earth's continental crust and oceans existed and were undergoing changes much earlier than previously thought (Harrison 2009). As a result, the conditions on Earth during the Hadean Eon may have been more suitable for the formation of early life rather than a purely inhospitable period, although the geological record of this time remains sparse.

Archean Eon

4.031 to 2.5 billion years ago

The Archean Eon was a time of significant biological and geological developments on Earth as the planet cooled and stabilized. Some of the first single-celled organisms appeared with the formation of the first supercontinent, Vaalbara, between 3.6 and 2.8 billion years ago. The name Archean is derived from the Greek word *Arkhaios,* meaning ancient. The Archean Eon is further divided on the International Chronostratigraphic Chart into four major eras: the Eoarchean (4 to 3.6 billion years ago), Paleoarchean (3.6 to 3.2 billion years ago), Mesoarchean (3.2 to 2.8 billion years ago), and Neoarchean (2.8 to 2.5 billion years ago). Due to the limited geological evidence available from this time period, as the early Earth's crust was still subject to constant recycling and metamorphism, there is a sparse rock record from the Archean Eon compared to later time periods. It is for this reason that the Archean sub-eras are not frequently used, and they will not be further explored in this book for geological and archaeological purposes.

Earth's interior was still very hot during the Archean Eon following the intense volcanic activity of the Hadean Eon, leading to faster mantle convection and active plate tectonics. Geological activity during the Archean Eon led to the formation of greenstone belts, distinctive zones of predominantly volcanic and sedimentary rocks that are commonly found within cratons. Cratons are stable and undeformed regions of continental crust that form the core of continents, preserving some of the oldest continental crust rocks that formed during the Archean Eon. Rocks within the greenstone belts have often undergone intense processes of metamorphism, which lead to their characteristic green color. The ancient basements of cratons are typically composed of metamorphosed rocks, including granulite-gneiss terrains. The composition of greenstone belts offers valuable insights into the tectonic processes and crustal evolution of continents.

Life was primarily microbial during the Archean, with the oldest bacterial fossils, known as stromatolites, dating to 3.5 billion years ago (Baumgartner et al. 2019). Earth's atmosphere was rich in gases such as methane, nitrogen, and carbon dioxide, which created oxygen-reduced (anoxic) conditions with negligible free oxygen. Biochemist Alexander Oparin proposed in his 1924 book *The Origin of Life* that life on Earth arose from inorganic compounds through the process of chemical evolution in Earth's early reducing atmosphere (Oparin 1938). In 1929, biologist John Haldane independently developed and published similar ideas about the early formations of life in a reducing atmosphere, speculating that organic molecules could have formed from inorganic materials (Tirard 2017). Both researchers focused on the concept of a "primordial soup," now known as the Oparin-Haldane hypothesis, where organic molecules formed in early oceans from the accumulation and interaction of inorganic compounds like water, ammonia, hydrogen, and methane. These compounds culminated in creating more complex structures such as lipids, proteins, and amino acids, which are basic building blocks for life to emerge.

The Oparin-Haldrane hypothesis has become a cornerstone in understanding the origins of life on Earth. In 1953, it was further supported by the Miller-Urey experiment, which simulated the early conditions of Earth to investigate the possibility of life arising from the interaction of inorganic compounds. The experiment used a closed-system apparatus, with one flask representing the early Earth's oceans, which was heated to mimic evaporation by creating water vapor. An electrical discharge of sparks was also generated to simulate lightning, providing the ambient charge necessary to initiate chemical reactions. This combination of processes recreated under early atmospheric conditions on Earth resulted in the formation of amino acids, including glycine and alanine. The experiment demonstrated the potential for abiotic synthesis during the Archean Eon; however, it is not without limitations. The exact composition of Earth's early atmosphere is still debated, and the mixture of gases used in the experiment may not have been entirely representative of its composition.

The end of the Archean Eon was marked by the Great Oxygenation Event (GOE), where cyanobacteria began to photosynthesize carbon dioxide and water into energy, releasing large quantities of oxygen into the atmosphere. This dramatically altered Earth's environment, paving the way for the evolution of more complex life to emerge, while simultaneously triggering a mass extinction of anaerobic bacteria, which could not tolerate the increasing oxygen levels. It also led to an increase in mineral diversity, with over half of the minerals on Earth today formed as a result of the GOE. Other hypothetical supercontinents believed to have emerged by the end of the Archean were Ur, which formed around 3 billion years ago, and Kenorland, which formed around 2.7 billion years ago.

Proterozoic Eon

2.5 to 0.541 billion years ago

The Proterozoic Eon was a time when life diversified rapidly after the appearance of the first multicellular organisms. The name Proterozoic is derived from the Greek words *protos*, meaning first, and *zoe*, meaning life. This eon is more uncommonly further divided into chronostratigraphic eras, namely the Paleoproterozoic (2.5 to 1.6 billion years ago), Mesoproterozoic (1.6 to 1 billion years ago), and Neoproterozoic (1 to 0.541 billion years ago). These are further subdivided into the following periods: Siderian, Rhyacian, Orosirian, Statherian, Calymmian, Ectasian, Stentian, Tonian, Cryogenian, and Ediacaran periods.

The large quantity of oxygen released during the photosynthesis of carbon dioxide and water by cyanobacteria that emerged in the Archean reacted with dissolved iron in the oceans during the Great Oxygenation Event (GOE). Sprawling banded iron formations resulted from this process, which has contributed to many of the iron sources on Earth today. As iron became oxidized, oxygen began to accumulate in Earth's oceans, having subsequent and dramatic consequences in the lower and upper atmosphere. The increasing oxygen levels were toxic to most anaerobic lifeforms during this time, resulting in a period of their mass extinction. Organisms that survived this extinction event could either already tolerate oxygen at this maximum or maintain survivability in anaerobic niches where the environment supported a low-oxygenated system. As the environmental pressure of oxygen toxicity changed Earth's ecosystems, it also drove the adaptation of aerobic metabolisms in lifeforms, which is a far more efficient process of producing energy (ATP).

The first well-known ice age, the Huronian Ice Age, occurred during the Paleoproterozoic Era approximately 2.4 to 2.1 billion years ago. This was a result of the GOE, which led to a significant increase in atmospheric oxygen concentrations and a decrease in greenhouse gases, such as methane, as it reacted with oxygen to form carbon dioxide and water as a by-product, reducing the greenhouse effect and triggering global cooling. The Huronian glaciation was one of the longest and most severe ice ages in Earth's history. Around the same time, the supercontinent Kenorland began breaking apart, slamming fragments into earlier landmasses, which facilitated the formation of the supercontinent Columbia (also referred to as Nuna) between 2.1 and 1.8 billion years ago.

By about 1.2 billion years ago, the continents of Laurentia (North America), Baltica (Northwestern Europe), Amazonia (parts of South America), Australia, and parts of both Africa and Antarctica began converging to form the supercontinent Rodinia. This resulted from the collision of fragments from the supercontinent Columbia (Nuna), which began breaking apart between 1.6 and 1.2 billion years ago. Between

750 and 600 million years ago, the supercontinent of Rodinia broke apart, and complex ecosystems subsequently began to develop as the supercontinent Pannotia formed from these fragments during the Cryogenian Period in the Neoproterozoic.

The Cryogenian Period, which lasted from approximately 720 to 635 million years ago, is noted for experiencing some of Earth's most intense ice ages, including the Sturtian and Marinoan glaciations. These are often referred to as "Snowball Earth" events, similar to that of the former Huronian glaciation, in which glaciers potentially covered the entire Earth, including the equator. This was followed by the Ediacaran Period, which marks a transition between the Cryogenian and Cambrian periods. The first macroscopic, multicellular animals known as metazoans emerged during this time as the Earth's oceans became fully oxygenated following the end of the Marinoan glaciation. As eukaryotes rose to prominence, some began to develop sexual reproduction, which was a major evolutionary milestone on Earth. By the final period of the Proterozoic, the Ediacaran, a variety of soft-bodied multicellular lifeforms came to dominate the planet. It is here that we find some of the earliest fossil evidence of lifeforms that are larger than single-celled organisms, building on a much longer evolutionary history that began considerably earlier (Knoll et al. 2006; Tang et al. 2024). This increase in biological diversity during the Early Ediacaran began to bolster the oxygen-rich atmosphere and stabilize environmental conditions on Earth, leading to a rapid onset of an evolutionary event known as the Cambrian explosion, marking the transition to the Phanerozoic Eon.

These eras and periods of the Proterozoic Eon will not be majorly focused on here, as it is a period where fossilization is less common, with a relatively sparse fossil record compared to the Phanerozoic Eon. While the Proterozoic Eon holds significant geological importance for understanding Earth's evolution and the development of early life leading to the Cambrian explosion, a large portion of the eon is covered by a majority of inaccessible rock, further limiting our ability to study it in detail.

Phanerozoic Eon

541 million years ago to present

The Phanerozoic Eon is considered the eon of visible life, and its beginning is marked by the Cambrian event, also known as the Cambrian explosion, which was a period when all major groups of plants and animals emerged rapidly. Life on Earth before the Cambrian explosion was primarily composed of single-celled and simple multicellular organisms. The name Phanerozoic is derived from the Greek words *phaneros*, meaning visible, and *zoe*, meaning life. This is the current eon in which we live and is further divided into the Paleozoic, Mesozoic, and Cenozoic eras, all of which will be covered in the following sections of this book, along with their associated geological periods. The abundant fossil record from this eon and its relatively recent age provide valuable insights into the history and evolution of life closer to Earth's outer surface.

As the Proterozoic Eon ended with an increase in biological diversity and the stabilization of ecosystems, many phyla of animals began to appear in the fossil record as time progressed through the multiple eras and their divisional periods of the Phanerozoic. These geological time distinctions represent some of the most recent major changes on Earth, including climatic shifts, biological diversification reflected in the fossil record, marine transgression and regression, tectonic movement, and other significant geological events. The supercontinent Pangea also formed during this eon as a result of continental collisions and subduction with fragments from Pannotia, which broke apart around 550 million years ago. These large landmass fragments consisted of Gondwana, Laurentia, Siberia, and Baltica, which collided to form Pangea. Pangea began formation approximately 335 million years ago, becoming fully assembled during the Paleozoic Carboniferous period 300 million years ago. This supercontinent persisted for nearly 100 million years before breaking apart around 200 million years ago, paving the way for the continents to drift to their present-day positions.

The following sections in this chapter will cover the three eras pertinent to the Phanerozoic Eon, beginning with the Paleozoic, Mesozoic, and Cenozoic, and continuing through all periods of each era. All major epochs of the Cenozoic Era periods will then be covered up until the recent Pleistocene and Holocene epochs. Subdivisions of epochs during the Paleozoic and Mesozoic eras are not as widely known as those of the Cenozoic Era due to the less detailed fossil record from these earlier time periods, making finer subdivisions of geological time difficult. For this reason, geologists often broadly refer to these periods using "Early," "Middle," and "Late" designations to represent these distinctions.

POST-CAMBRIAN EVENT

These sections are primarily set up for the reader to gain a general understanding of the recent geological events that influenced Earth's climate and tectonic movement, which in turn had overarching consequences for life on the planet. As time nears the Neogene Period of the Cenozoic Era, the reader will be taken through the non-linear sequence of mammalian evolution and its lineage progression to early primates, and later, to early hominins and *Homo sapiens*. The chapter that follows will cover hominin evolution and the geoarchaeological evidence for their incorporation into our human ancestry, which will transition into the earliest evidence of the Stone Age in the Prehistoric Ages, marked by the beginning of stone tool use by early hominins.

Paleozoic Era

541 to 252 million years ago

The Paleozoic Era is the first era of the Phanerozoic Eon. It is also known as the Age of Ancient Life for the emergence of the first fish, amphibians, and invertebrates during the Cambrian Period. It is divided chronologically into the Cambrian, Ordovician, Silurian, Devonian, Carboniferous (Mississippian and Pennsylvanian), and Permian periods. The name Paleozoic is derived from the Greek words *palaios*, meaning ancient, and *zoe*, meaning life. Land plants began to diversify as a result of the supercontinent Pannotia breaking apart into the continents of Gondwana and Laurasia, along with some smaller land masses, which later formed the supercontinent Pangea by the Late Carboniferous Period.

This era was significant in that the colonization of Earth was not only attributed to animals adapting to life from the oceans, but also of plants and trees beginning to take root on major continental landmasses. While the Cambrian explosion is often cited as integral to the rapid global biodiversity that appeared during this time on Earth, the eruptive rate of evolution caused many species to meet extinction quickly, with short genus longevity during the Cambrian (Kröger, Franeck, and Rasmussen 2019). As ecosystem disruptions continued to stabilize during the following Ordovician and Silurian periods, this was met with a greater chance of persistence among lifeform genera. For many lineages, these periods were crucial for the appearance of ancestral species that diversified and expanded from existing phyla, with the rise of new life forms within those established groups. This is especially the case during the Great Ordovician Bidiversification Event (GOBE), where the greatest increase of marine biodiversity in Earth's history occurred before a subsequent mass extinction event at the end of the Ordovician Period.

As life began to move from the oceans to adaptations on land during the Silurian and Devonian, the earliest group of synapsids, formerly known as pelycosaurs, appeared later during the Late Carboniferous Period. They are now more directly referred to as non-mammalian or basal synapsids, characterized as primitive ancestors to later therapsids. By the end of the Permian Period, the divergence from the main synapsid lineage would give rise to the first proto-mammals known as therapsids. These are a group of synapsids that includes mammals and their ancestors, and became the dominant large land animals in the latter half of the Permian Period. Despite most therapsids going extinct by the end of the Triassic Period, one lineage known as cynodonts persisted and became the ancestor of mammals. These animals possessed distinct mammalian traits, including being warm-blooded and likely developing fur, while retaining some characteristics found in reptiles, such as the ability to reproduce by laying eggs.

The Paleozoic was a crucial time for the development of life on Earth, from the eruption of lifeform diversity during the Cambrian Period to the earliest ancestors of mammals in the Permian Period. It also laid the groundwork for the emergence of mammals and their continued survival alongside the rise of dinosaurs on Earth during the Mesozoic. While mammals initially remained in smaller ecological niches throughout the first two eras of the Phanerozoic, their persistence would lead to the rise of primates in the Cenozoic Era.

Mesozoic Era

252 to 66 million years ago

The Mesozoic Era is the second era of the Phanerozoic Eon. The name Mesozoic is derived from the Greek words *mesos*, meaning middle, and *zoe*, meaning life. This era is divided chronologically into the Triassic, Jurassic, and Cretaceous periods, each of which will be discussed chronologically in later sections. The Mesozoic as a whole is primarily known for the rise and subsequent fall of dinosaurs, which was marked by the K-Pg event (Cretaceous-Paleogene extinction event). As reptiles became the dominant land species on Earth throughout the Mesozoic Era, the first birds and true mammals also began to appear. Notably, these first true mammals emerged from synapsids of the Late Paleozoic Era. The emergence of these true mammals in the fossil record is covered throughout the Permian and Triassic Period sections, where the diversification of species can be examined within more constrained references of time.

A major geological event at the beginning of the Mesozoic Era, specifically the Late Triassic Period, was the reshaping of the continents. Initially in the Early Mesozoic, dinosaurs were partitioned by swaths of temperature zones in the upper latitudes of Pangea, where vast xeric deserts created geological barriers, intense periods of heat, and drought in the tropics. The lack of vegetation in these equatorial regions initially restricted the dispersal of herbivores due to a lack of stable subsistence. Pangea began to break apart into Laurasia and Gondwana by the Late Triassic to Early Jurassic periods, gradually leading to the continents we observe today and facilitating the dominance of dinosaurs on the planet during the Mesozoic.

The breakup of Pangea resulted in ocean current reorganizations, which contributed to a warm and humid global climate during this era. The warmer climate of the Mesozoic contributed to higher sea levels, leading to the widespread inundation of continental landmasses, creating extensive shallow inland seas. These vast warm seas were home to marine reptiles, including ichthyosaurs and plesiosaurs, which emerged as separate lineages alongside dinosaurs in the Triassic Period. Inland seas particularly became expansive during the Late Jurassic and Cretaceous periods, where marine reptiles like mosasaurs, rather than their predecessors of the Triassic, were the apex predators of these ocean ecosystems. One of these vast seas, the Western Interior Seaway, divided North America into two landmasses during the Late Cretaceous Period: Laramidia and Appalachia. At its most extensive point, the Western Interior Seaway reached nearly 2,000 miles long and 600 miles wide, stretching from the Arctic Ocean down to the Gulf. By this time in the Late Mesozoic, ichthyosaurs had nearly gone extinct, while plesiosaurs were experiencing a decline in population.

By the end of the Mesozoic Era, the nearly 165 million-year reign of dinosaurs on the planet would come to an end. The ultimate and immediate demise of the dinosaurs on Earth is attributed to an asteroid impact on the Yucatan Peninsula around 66 million years ago, creating the Chicxulub crater, which will be discussed in further detail within the section covering the Cretaceous Period. This impact triggered a global catastrophe as sunlight became eclipsed by a cloud of ash, leading to the extinction of all non-avian dinosaurs on Earth. However, the end of reptilian dominion opened the door for mammals to diversify rapidly after the K-Pg event, which would consequently set the stage for the emergence of early primates in the Cenozoic Era.

Cenozoic Era

66 million years ago to present

The Cenozoic Era is the third era of the Phanerozoic Eon and is also the current geological era in which we live. The name Cenozoic is derived from the Greek words *kainos*, meaning new, and *zoe*, meaning life. All of human history, and prehistory, occurred during this geological sequence. It is divided chronologically into the Paleogene, Neogene, and Quaternary Periods, each with its own respective epochs. These epochs are often defined in the geological record based on periods of deposition, erosion, and stability reflected in stratigraphic sequences. All of the epochs within the Cenozoic Era will be discussed in the sections following the geological periods of the Phanerozoic Eon.

The extinction of non-avian dinosaurs in the Mesozoic Era created ecological vacancies that mammals rapidly adapted to and diversified out of. Early primates and modern ungulates (hooved animals) emerged, which reshaped the food webs of spreading grassland ecosystems. The lineage of early hominins began to diverge from the lineage leading to chimpanzees and bonobos around 6 to 8 million years ago, which was a major step in the evolutionary journey that ultimately led to the emergence of *Homo sapiens*. For these reasons, the Mesozoic is often referred to as the Age of Mammals.

Earth underwent multiple glacial and interglacial cycles throughout the Cenozoic, particularly in the Quaternary Period. The most recent ice age and glacial period began around 2.58 million years ago, marking the start of the Pleistocene Epoch. Around 26,500 years ago, the ice sheets on Earth reached their fullest extent, known as the Last Glacial Maximum (LGM), which peaked around 20,000 years ago. By 19,000 years ago, the LGM ended as ice sheets began to melt and sea levels rose eustatically over the next several thousand years. Around 12,900 years ago, the final cold snap of the Last Glacial Period occurred, known as the Younger Dryas, before Earth entered its current interglacial period around 11,700 years ago. This would mark the beginning of the Holocene and the rise of agricultural development facilitated by the post-glacial climate.

Cambrian Period

541 to 485.4 million years ago

The Cambrian Period is characterized by the Cambrian explosion, which was a rapid diversification of life as a result of complex environmental and biological factors. It is often attributed to increasing oxygen levels and tectonic activity coinciding with the end of a glacial period. The tectonic activity caused by the breakup of the supercontinent Pannotia led to the expansion of ocean basins and shaped many ocean currents. This had a significant impact on Earth's geology and climate in the Cambrian Period. A rise in sea levels from melting glaciers also formed vast warm and shallow seas, which provided ecosystems with abundant nutrient availability for new marine life. This explosion of life marked the appearance of most major animal phyla in the fossil record, with many species developing hard exoskeletons and shells to fill new ecological niches as a result of changing environmental pressures.

Trilobites also became one of the most common animals during this time, along with the diversification of many other early marine invertebrates and arthropods. This in turn marked a significant change in the fossil record where animals with hard shells became frequently fossilized. While there are periods of glaciation that precede and follow the Cambrian, this time on Earth was generally much warmer, with less ice formation and no major glaciations. As a result, most complex life during the Cambrian Period emerged in the oceans, as trees and grasslands had yet to develop on land. While some of the earliest animal life existed in the preceding Ediacaran Period of the Proterozoic Eon, the Cambrian marks the point where many major animal phyla emerged in an explosive diversification that built upon previously existing biota.

Ordovician Period

485.4 to 443.8 million years ago

The Ordovician Period was a time of significant diversification of marine life and the emergence of early land-adapted plants. The warm and wet climate of the Early to Middle Ordovician allowed new groups of marine invertebrates to emerge, including bryozoans and brachiopods. Fish also became more abundant following their appearance in the Cambrian explosion. Earth experienced another glaciation period during this time, which was triggered by the supercontinent Gondwana moving across the South Pole. This event, the Hirnantian glaciation, occurred during the Late Ordovician Period roughly 445 to 443 million years ago. It became a catalyst for what has come to be known as the first of the "big five" most severe and consequential mass extinction events in Earth's history.

The corresponding drop in sea level due to glacial formation at the end of the Ordovician Period initiated the Ordovician-Silurian Extinction Event, also known as the Late Ordovician mass extinction (LOME). Much of the marine life that thrived in warm shallow seas became impacted by the significant drop in temperature and lower sea levels. This wiped out a significant portion of marine life due to the altered environmental conditions. The LOME is considered one of the more significant mass extinction events in Earth's history, estimated to have eliminated 85% of marine species existing at the time. It has been found that subsequent anoxia and warming triggered by a pulse of resumed volcanic activity on Earth about one million years later may have also contributed to the mass extinction event (Bond and Grasby 2020). Large igneous provinces (LIPs) erupted towards the end of the Ordovician, which released greenhouse gases back into the atmosphere and toxic metals such as mercury into the ocean.

The warming trend that followed as a result led to the expansion of oxygen-minimum zones in oceans, potentially causing a second wave of extinction. For a brief period during the Middle to Late Ordovician, however, oxygenation resumed, although warm saline ocean currents continued to be a contributing factor to oxygen depletion along wide swaths of the deep seafloor. While this period of oxygenation expanded many ecosystems, oxygen-minimum zones were still in flux during the LOME and into the Silurian Period. The combination of glaciation and volcanism that led to such a dramatic decline in marine life has shifted the understanding of specific roles these events played by providing an example of subsequent geological events that can trigger periods of extinction on Earth.

Silurian Period

443.8 to 419.2 million years ago

The Silurian Period was a time of relatively warm, stable climates with rising sea levels and significant geological activity following the major glaciation of the Ordovician Period as ice sheets began to melt. Water that was once locked in glacial ice contributed to sea-level rise and the inundation of continental shelves, creating warm shallow lacustrine and marine environments, which became filled with distinct postglacial deposits. Tectonic plate movement caused the formation of volcanic island chains and a continued shift in the continent Gondwana throughout the period, which formed as a result of Pannotia breaking apart. The Silurian is characterized by a shift to life on land by the first terrestrial arthropods. The first jawed species of fish, known as acanthodians, also appeared as fish continued to become increasingly prominent. This period ended with several minor extinction events linked to climate change.

Early land plants began diversifying as well, with vascular plants that exhibit rigid stems emerging in order to adapt to local climatic and paleogeographic locations across different landmasses. Trilete spores, a type of spore found in ferns, mosses, and other early land plants, from the Early Silurian and Late Ordovician indicate that terrestrial flora had developed strategic evolutionary adaptations for resilience and widespread dispersal during the Silurian (Pšenička et al. 2021). During the Late Silurian, giant, tree-like fungi known as *Prototaxites* dominated parts of Earth's landscape and persisted into the Late Devonian Period. While lacking a complex vascular system, they still formed large trunk-like structures composed of interwoven vascular tubules and reached heights of up to eight meters. *Prototaxites* would have been adapted to environments with high water availability, such as river valleys, rather than widespread distribution across every terrestrial paleolandscape. There is an ongoing debate about the classification of these organisms as fungi, with some researchers pointing to the fact that fossil evidence indicates a lack of chitin in *Prototaxites*, like that of other extinct and extant species of fungi that exhibit this feature in their cell walls (Loron et al. 2025). Thus, the conclusion is that *Prototaxites* represent an extinct lineage of multicellular terrestrial organisms that contributed to a wider lost ecosystem of life which dominated Earth before the rise of large trees. Tree-like plants would replace *Prototaxites* in the Late Devonian and continue to flourish into the Carboniferous Period, where vast swamps and low-lying tropical regions created oxygen-reduced backwater environments, and decaying plant remains were waterlogged and preserved. These layers of peat became buried under sediment and subjected to immense heat and pressure over millions of years of tectonic movement, creating vast coal deposits found on Earth today.

Devonian Period

419.2 to 358.9 million years ago

The Devonian Period is often referred to as the Age of Fishes, as many species of fish diversified rapidly during this time. Early jawed marine vertebrates, which emerged in the Late Silurian Period, continued to diversify with armored species of fish (placoderms) such as *Dunkleosteus* emerging in the Late Devonian. Fossil evidence of an early shark-like fish known as *Doliodus problematicus* suggests a transition between jawed fish, known as acanthodians, and modern sharks during the Early Devonian. During the Late Silurian to Early Devonian periods, the species subclass Ammonoidea emerged in the fossil record. Ammonoidea includes the group of the first true ammonites, Ammonitida, which existed in the Jurassic Period. The first amphibians also appeared alongside tetrapods, marking a significant transition in life from water to adapting for survival on land.

Land plants began to diversify across widespread areas, including the appearance of the first forests following the decline of *Prototaxites*. As vascular plants continued to diversify and develop features like root structures and stronger cell wall support, they became increasingly efficient at capturing sunlight and extracting nutrients from the soil. This likely led to competition with *Prototaxites*, eventually leading to their complete replacement by forested ecosystems in the Carboniferous Period. However, the vascular plant fossil record from the Silurian to Devonian periods shows a significant sampling bias distorting global diversity patterns and is influenced by the best-sampled continents. Spatial differences in flora species have revealed that geographical and climate barriers controlled the distribution of terrestrial vegetation (Capel et al. 2022). Accounting for this bias, a corrected diversity pattern reflects an increase of vascular plants potentially alongside *Prototaxites* in Early to Middle Devonian paleolandscapes.

Tectonic activity during the Devonian led to the collision of many landmasses and the uplift of mountain ranges. Through this process of mountain-building and increased volcanic activity, a cascade of environmental changes began to transform the planet. The expansion of terrestrially adapted plants altered global carbon levels, which contributed to the growth of polar caps, transitioning Earth's climate from a greenhouse to an icehouse climate. Tectonic collisions altered landscapes by closing oceans, raising sea levels, and disrupting nutrient cycling and deep ocean currents. Vascular land plants that were developing new root systems broke down bedrock and created deep soils, which contributed to nutrient and phosphorus runoff into the oceans, triggering algal blooms through a process known as eutrophication. Anoxic conditions once again began to affect deep oceans as a result, which triggered the second of the "big five" mass extinctions known in Earth's history, the Late Devonian mass extinction, between 372 and 360 million years ago (Smart et al. 2023). This extinction event came in two pulses, with approximately 85-70% of all marine species suffering extinction, including many species of trilobites and all species of placoderms. Marine ecosystems were placed under significant stress following the impact on plankton, sponges, and coral in the oceans.

Carboniferous Period

358.9 to 298.9 million years ago

The Carboniferous Period is known for contributing to the abundant coal deposits on Earth today and widespread limestone formations due to the extensive amount of swamp environments present during this time. Low-lying tropical regions created oxygen-reduced backwater environments, which preserved decaying plant remains in waterlogged conditions. Coal deposits were formed from layers of peat that became buried under sediment and subjected to immense heat and pressure over millions of years of tectonic movement. The Carboniferous Period is commonly divided into the Mississippian and Pennsylvanian subperiods in North America due to geological distinctions of different rock formations.

Vascular plants and forested regions sprawled across vast landscapes and began to transform other ecosystems as well, while also contributing to significantly increased levels of oxygen in Earth's atmosphere compared to those observed today. The Carboniferous is perhaps most widely recognized as a geological time period for the appearance of giant predatory insects, such as *Meganeura* (Nel et al. 2018), due to the increased atmospheric levels of oxygen. The collision of Gondwana with Laurasia, forming the supercontinent of Pangea, was a major geological event occurring during the Carboniferous Period. This created drastic changes in Earth's climate, which affected ocean currents and led to Earth's cooling along with the formation of glaciers.

The Carboniferous Period was also considered the Golden Age of Sharks, as many prehistoric shark species emerged, including *Stethacanthus*, *Helicoprion*, and *Falcatus*. It was also a time of diversification for amphibians, giant arthropods, and the first appearance of reptiles. The first pelycosaurs emerged during the late Carboniferous, around 320 million years ago, and are the earliest known synapsids, which are the group that includes mammals and their extinct relatives. These basal (primitive) synapsids were once referred to as mammal-like reptiles; however, as they are not reptiles and are not included in reptile lineages, the term stem mammal has since replaced this terminology to include the entire synapsid lineage. It has also since replaced the term pelycosaur as this only represents a common ancestor and not therapsids as descendants (paraphyletic), and consequently excludes mammals, which are descendants of therapsids. These basal synapsids possess characteristics that represent the early branching point for the mammalian lineage as precursors that paved the evolutionary path towards the emergence of proto-mammals.

PALEOZOIC ERA

Permian Period

298.9 to 251.9 million years ago

The Permian Period was a time when Earth experienced drier climates and widespread glaciation as Pangea finalized its formation. The first true ancestors of mammals, known as cynodonts, emerged toward the end of the Permian Period. They are a specific group of shrew-like creatures called therapsids that are, in turn, a subgroup of the broader synapsid clade. This period ended with the largest mass extinction event in Earth's history, the Permian-Triassic extinction event, which was the third of the "big five" extinctions on Earth that killed off about 90% of all marine life and 70% of terrestrial vertebrate species.

The cynodont species' extensive diversification after the Permian-Triassic extinction event gave rise to the first true mammals in the Early to Middle Mesozoic Era. The earliest known cynodont, known as *Charassognathus gracilis*, was discovered in South Africa and dates back to around 259.5 to 254.1 million years ago (Botha, Abdala, and Smith 2007). The discovery of *Charassognathus gracilis* helps fill a gap in the fossil record by tracing the evolution of mammalian characteristics, where the divergence between the synapsid lineage leading to mammals and the sauropsid lineage leading to reptiles occurred.

Triassic Period

251.9 to 201.3 million years ago

The Triassic Period is the first period of the Mesozoic Era, and was a time of recuperation for many species from the Permian-Triassic extinction event. Pangea was the sole supercontinent on Earth, creating hot and dry conditions on the interior and seasonal monsoons along some coastlines. Dinosaurs began to appear during this period, along with the first true mammals. These early true mammals are generally considered to be the morganucodontids, such as *Eozostrodon*, which appeared during the Late Triassic Period around 210 million years ago.

While sharing similar characteristics with earlier cynodonts due to a shared common ancestor, they are not considered to have evolved directly from them, as the lower jawbone indicates a transition towards more mammalian-like features. Specifically, the Morganucodon is known for having a single lower jawbone, a defining feature of mammals where the posterior jawbones have moved to become part of the middle ear, and the lower jaw becomes isolated. More recent research suggests that *Brasilodon quadrangularis*, another shrew-like creature from the Late Triassic (225 million years ago), may have potentially been the first true mammal-like creature (Cabreira et al. 2022). This is due to the species exhibiting similar features of mammals in its dentition, indicating it may have been a diphyodont (possessing two sets of teeth in a lifetime).

Around 201 million years ago, tectonic activity in the Late Triassic brought the initial phases of Pangea's breakup, which continued over the next 150 million years until the continents drifted to their present-day positions. Volcanic eruptions in the Central Atlantic Magmatic Province (CAMP) caused by Pangea's rift released immense amounts of carbon dioxide into the atmosphere (Davies et al. 2017). Ocean anoxia and acidification again followed by climatic warming, killing off many marine invertebrates and large reptiles in the fourth most devastating extinction event in Earth's history, the End-Triassic mass extinction (Schoepfer et al. 2022). This mass extinction, the fourth of the "big five" in Earth's history, occurred in two pulses separated by several hundred thousand years, challenging the view of a single rapid extinction that led to the dominance of dinosaurs in the Jurassic Period. This is based on evidence from the British Isles, where a second mass extinction horizon is seen in the upper Langport Member of the Lilstock Formation, separated by a brief period of recovery and a resumed increase in biological diversity from the extinction horizon in the lower Cotham Member of the same formation (Wignall and Atkinson 2020). The End-Triassic Extinction enabled the rapid diversification of dinosaurs in the Jurassic Period after the elimination of competitors during these extinction pulses, opening up empty ecological niches to fill.

Jurassic Period

201.3 to 145 million years ago

The Jurassic Period is distinctly recognized for the dominance of dinosaurs on Earth, however, there is widespread misinterpretations about which dinosaurs roamed the planet during this time. To note one misconception, many associate the *Tyrannosaurus rex* with the Jurassic Period and living alongside *Stegosaurus*, which existed in the Late Jurassic, however, *T. rex* did not emerge until the Late Cretaceous Period, meaning *Stegosaurus* was extinct for nearly 76 million years before the first appearance of any *T. rex*. Rather, in the Jurassic Period, giant sauropods, such as *Brachiosaurus* and *Diplodocus*, thrived alongside other prey and armored herbivores like stegosaurs, which were hunted by carnivorous species like *Allosaurus* and *Dilophosaurus*, to name a few of the vast number of dinosaur genera that existed throughout the Jurassic. Due to the incomplete fossil record, it is still impossible to reconstruct the exact number of genera and species that existed simultaneously alongside one another at different stages of the Jurassic, let alone the wider encompassing Mesozoic Era.

One of the major geological changes during the Jurassic Period was the continued breakup of the supercontinent Pangea into Laurasia and Gondwana, opening the Atlantic and Indian Oceans. Over millions of years, these landmasses would slowly fragment, becoming the continents we observe today. Lush vegetation and warm, thriving ecosystems spanned the continents of Laurasia and Gondwana, which supported diverse flora and fauna. The earliest birds, known as *Archaeopteryx*, also appeared during the Jurassic Period, derived from a group of dinosaurs called theropods. Mammals filled a wide range of niches alongside dinosaurs, including burrowing, climbing, and nocturnal or semi-aquatic lifestyles and continued to undergo extensive diversification of different species into the Cretaceous Period. Marine reptiles like ichthyosaurs and plesiosaurs, as well as the first true ammonites known as Ammonitida, also emerged in the oceans. Alongside this marine life, the oldest known group of modern sharks, *Hexanchiformes*, evolved around 195 million years ago. Shallow seas associated with lagoons and coastal areas covering large areas contributed to the overall preservation of marine fossils from this time period. Inundation quickly buried fossils in lacustrine and marine sediments during periods of transgression, protecting them from weathering and erosion throughout the next several million years.

Cretaceous Period

145 to 66 million years ago

The Cretaceous Period was the final period of the Mesozoic Era, and saw the rise of flowering and fruit-bearing land plants and the appearance of the first grasses, which contributed to new food sources in ecosystems. Throughout the Cretaceous, Laurasia and Gondwana continued to break apart after splitting from Pangea, forming new ocean basins, notably the Atlantic and Indian oceans. This in turn fueled higher sea levels that inundated continental shelves, along with creating a warmer greenhouse climate, which caused oceanic anoxic events (OAEs) (Scotese et al. 2024). Between approximately 125 and 80 million years ago, the rise of flowering land plants (angiosperms) drove a coevolution with pollinating insects as new ecological niches emerged, known as the Cretaceous Terrestrial Revolution (KTR). This created opportunities for small mammals, snakes, birds, and lizards to diversify alongside dinosaurs.

There has been much debate over the driving factors of the KTR and its impact on the adaptability of dinosaurs. A hypothetical rapid surge in their diversity during the Middle to Late Cretaceous has been criticized and attributed to a bias in fossil sampling rather than true evolutionary pulse which was driven entirely by the KTR; however, earlier studies found that dinosaurs were not in decline leading up to the end of the period suggesting that this diversification had been ongoing (Lloyd et al. 2008). A more recent study challenging this notion found that while increased dinosaur diversity throughout the Middle to Late Cretaceous can be attributed to better fossil sampling from this period, dinosaurs were not central to the KTR and, in effect, could have already been experiencing a widespread decline (Benton 2023). However, considering this has been a long ongoing controversy in paleontology of the Cretaceous, arguments have been made against this claim since its publication. Researchers have noted factors such as a lack of a complete geological record from the Cretaceous (Dean et al. 2025). Models of the Cretaceous paleoecology in western North America have also shown that, contrary to previous ideas of a gradual decline up until the K-Pg, dinosaurs were flourishing in places like New Mexico during the Late Cretaceous (Flynn et al. 2025).

Perhaps of most significance is the dramatic end to the Cretaceous and Mesozoic as a whole, ushering in the dawn of mammalian domination on Earth following the last of the "big five" extinction events. This period ended with the second-largest mass extinction event in Earth's history, known as the Cretaceous-Paleogene Extinction Event (K-Pg), which eliminated all non-avian dinosaurs, leaving small ground-dwelling mammals, earlier species of birds, and smaller deep-water species of sharks. This created the K-Pg boundary, a thin layer of marker sediment found globally that contains high levels of iridium, one of the rarer elements in Earth's crust, yet found abundantly on asteroids and comets. A distinct transition in the fossil record can be observed at this boundary, where dinosaurs and other Mesozoic organisms are absent above it. Strong evidence suggests that the K-Pg extinction event coincided with a massive asteroid impacting the Yucatan Peninsula in Mexico approximately 66 million years ago, creating the Chicxulub crater measuring 180 to 200 kilometers in diameter (Chiarenza et al. 2020; Gulick 2024). The asteroid is estimated to have been between 10 and 15 kilometers in diameter itself and released the equivalent of nearly 100 teratons of TNT, roughly 4.5 billion times the force of the Hiroshima atomic bomb.

Mammals underwent selective pressures following the initial impact winter, which likely inhibited photosynthesis for a brief period of time and devastated subsistence strategies. Overall cooling may have lasted up to a decade or more as soot, silicate dust, and sulfur significantly hindered sunlight from reaching Earth's surface. However, the K-Pg event likely vacated ecological and predatory roles that dinosaurs once filled, and the previous diversification of mammals throughout the KTR largely contributed to their survival following the asteroid impact. Tectonic uplift created new habitats and landscapes during the KTR while both contributed to climate heterogeneity, further supporting mammalian diversification (Weaver et al. 2024). The idea that mammals experienced a delayed rise in the

Eocene following the K-Pg event has largely been debated (Meredith et al. 2011; Bininda-Emonds and Purvis 2012; Murphy et al. 2012), though there is little doubt that the extinction of the dinosaurs did not open new avenues for mammals to begin adapting to a new world, one in which they would come to dominate.

Paleogene Period

66 to 23.03 million years ago

The Paleogene Period is characterized by the rapid diversification of birds and mammals during this time of recuperation after the Cretaceous-Paleogene Extinction Event (K-Pg). It is divided chronologically into three epochs: the Paleocene, Eocene, and Oligocene. The use of epochs, particularly in the Cenozoic Era periods, is more commonly used than in the Paleozoic and Mesozoic eras due to the more recent and detailed nature of the geological and fossil records of the Cenozoic Era. Epochs are useful for studying the rapid diversification of life by tracking more specific events that occurred within a shorter time frame, reflected in the geological record. Initially, the Paleogene Period as a whole began with a warmer, moist climate in its first epoch, the Paleocene. Shortly thereafter, Earth's climate began to trend towards significantly cooler temperatures in the Eocene and Oligocene, leading to the formation of glaciers and polar caps, which stimulated the expansion of emerging grasslands.

Following the events of the Cretaceous Period and into the current Cenozoic Era, the Paleogene truly began the long-standing, modern order of mammalian evolution on Earth. The dramatic and literal impact that the K-Pg caused to ecological niches left ecosystems disordered for some time on the planet.

Mammals began to fill these predatory vacancies, left behind by the now-extinct dinosaurs, and came to dominate food chains as they emerged with larger and more varied forms during this period. This expansion of mammalian genera brought with it the rise of ancestral primates during the Paleocene, which is the earliest appearance of some of these distinct features observed in mammalian phyla.

The earliest ancestors of primates, known as plesiadapiforms, evolved from archaic mammals and exhibited primate-like features like grasping hands and feet. Among the earliest primate fossils are those of the genus *Purgatorius*, dating to within 139,000 to 105,000 years after the K-Pg Event, which encompasses several species of proto-primates known as stem primates (Wilson Mantilla et al. 2021). While not a direct ancestor of modern primates, characteristics of this genus represent a transition to the first true primates, as they were small, tree-dwelling, omnivorous mammals that excelled at numerically dominating arboreal ecosystems. The first fossils were discovered in the Hell Creek and Tullock formations in northeastern Montana, and the early dates retrieved from the holotypes revealed that *Purgatorius* likely began to develop in the Late Cretaceous, with the radiation of these early primates already underway by the Early Paleogene.

Neogene Period

23.03 to 2.58 million years ago

The Neogene Period is particularly distinct for the evolution of grazing mammals following the spread of grasslands as a consequence of the gradual cooling at the end of the former period, along with the emergence of early hominins in Africa. This period is divided chronologically into two epochs: the Miocene and Pliocene. As Earth's climate started to cool during the preceding Oligocene Epoch, the last epoch of the Paleogene Period, the global expansion of grasslands began to replace many forests and create new ecosystems. The Neogene Period saw the rise of upright, bipedal locomotion in early hominins as a result, which began to decrease the frequency of their arboreal lifestyles. According to the savanna hypothesis, this was driven by factors like predation, visibility, and long-distance migration, requiring adaptation to life on the ground. However, some early hominin fossils, like *Ardipithicus*, have been associated with tree or bush savanna habitats, suggesting that the rise of bipedalism in early hominins was not solely driven by the need to adapt to open grasslands (Cerling et al. 2010). Other species of great apes and prehistoric primates emerged earlier at the beginning of the Miocene, with genera of early hominins like *Ardipithecus* and *Australopithecus* appearing in the Late Miocene and Early Pliocene.

The existence of one of the first *Homo* species, *Homo habilis*, during the Late Pliocene is often debated, as more concrete evidence suggests they emerged in the Early Pleistocene (Kimbel, Johanson, and Rak 1998). However, the discovery of a jawbone with teeth at Ledi-Geraru in the Afar Region of Ethiopia, which is attributed to the genus *Homo*, has since challenged the timeframe of their first appearance, suggesting early *Homo* species may have existed as early as 2.8 to 2.75 million years ago during the Pliocene (Villmoare et al. 2015). When early hominins first appeared in this period, it set the stage for the entire course of human evolution, which includes the development of key adaptations such as bipedalism and the beginnings of stone tool use.

Marine ecosystems also underwent fluctuations throughout the Neogene Period as the Antarctic ice sheets stabilized and Northern Hemisphere glaciation intensified. These cooling trends influenced ocean circulation patterns, gradually leading to the glacial conditions of the Quaternary Period. Around 23 million years ago, just after the start of the Neogene Period, Earth's largest known shark species, *Otodus megalodon,* emerged in the Early Miocene and existed for roughly 20 million years until the Late Pliocene. While the Neogene Period was generally much cooler than the preceding Paleogene Period, the brief warming trend in the Middle Miocene provided favorable ecosystems for these large marine predators. Shallow inland seas created diverse coastal environments and supported a rich food web including whales, seals, sea turtles, and other marine mammals. The continued cooling trend in the Late Neogene, along with significant geological and oceanographic changes, such as the formation and rise of the Isthmus of Panama, altered ocean currents and contributed to a reduction in prey, primarily in whale populations, driving megalodons to extinction by the Pliocene Epoch. Their disappearance had cascading transoceanic effects on marine ecosystems as different species, such as great white sharks (*Carcharodon carcharias*) and orcas (*Orcinus citoniensis*), would come to fill the role these massive sharks once did, reshaping the food chain for marine mammals.

For human evolution, the Neogene imparts a critical role in shaping our contemporary understanding of our ancestral prehistory. Around the Late Miocene, we find some of the earliest anatomical evidence in early hominins towards incipient bipedalism and major dietary shifts. This critical period is believed to have been when the human lineage diverged from that of great apes. By the end of the Pliocene, the genus *Homo* emerges and begins to reshape the stone tool industries set in motion by hominins in Africa. This period of hominin evolution is covered in greater detail in the following sections on each epoch of this period, the Miocene and Pliocene, and in the next chapter on paleoanthropology by bringing the reader back to the Neogene Period.

Quaternary Period

2.58 million years ago to present

The Quaternary Period is divided chronologically into two epochs: the Pleistocene and Holocene. It is the period in which we currently reside. Earth has undergone significant climate changes during this time, including a series of roughly thirty glacial and interglacial cycles, which created new ocean and lake basins as massive ice sheets melted. It is the period in which modern *Homo sapiens* became the dominant species on Earth, beginning with the emergence of early hominins near the end of the Miocene Epoch. Glacial periods during the Pleistocene Epoch allowed for the formation of land bridges, which facilitated the movement of animals and humans across continents and into the Americas.

As glaciers advanced and retreated during the Quaternary Period, sea level fluctuations caused periods of stability, erosion, and deposition, which are reflected in the geological record. The glacial till, eolian, fluvial, volcanic, and alluvial sediment deposits that buried Quaternary paleosols reflect the separation between Pleistocene and Holocene deposits, with the latter representing the current warm period. Being the most recent geological strata, the Quaternary sediment and rock record has had less time for erosional processes to occur, leaving it relatively well-preserved and accessible.

Unlike older sediments, Quaternary deposits are generally loose and unconsolidated, having not yet undergone cementation into a hard rock layer. These represent recent environmental changes, including glacial activity, climate change, and landscape evolution. Floral and faunal fossil assemblages contained within these sediments can also share similar characteristics to modern-day species, providing clues about the extent of past environments and ancient ecosystems, otherwise known as paleoenvironments. Quaternary stratigraphy is the most accessible deposit on Earth, and thus, we are able to study recent changes in the fossil record with the best accuracy of all geological periods.

PALEOGENE PERIOD

Paleocene Epoch

66 to 56 million years ago

Beginning with the Paleogene Period epochs, within the larger encompassing Cenozoic Era, the Paleocene Epoch began following the extinction of the dinosaurs during the Cretaceous-Paleogene Extinction Event (K-Pg). As many species of mammals and birds underwent a period of recuperation after the K-Pg event, a species considered among the earliest ancestors of primates, *Purgatorius*, diverged from a lineage of archaic mammals about 65.9 million years ago (Wilson Mantilla et al. 2021). Although not directly ancestral to modern primates, they represented a distinct transition to some of the earliest true primates. This genus represents several species of extinct proto-primates, known as plesiadapiforms, which exhibited primate-like features like grasping hands and feet.

The Paleocene-Eocene Thermal Maximum (PETM) occurred near the end of the period that saw Earth's temperatures trend toward a warmer climate (Tierney et al. 2022). This was likely triggered by volcanic activity in the North Atlantic Igneous Province (NAIP), contributing to a release of excess greenhouse gases. Analysis of sedimentary cores from the North Sea provides evidence of increased levels of mercury relative to organic carbon, which is a proxy for volcanic emissions (Kender et al. 2021). The warming climate weakened ocean circulation, hindering the delivery of oxygen to deeper waters, which led to widespread deoxygenation and acidification.

This resulted in a mass extinction of marine life, with between 35-50% of deep-sea benthic foraminifera species going extinct (Foster et al. 2013; Alegret et al. 2021). Although a significant extinction event, it simultaneously triggered adaptations and diversification of life in surface oceans and on land with rapid species turnover. By around 55 million years ago, modern primates, early horses, and the ancestors of pigs, whales, cows, and hippopotami emerged, coinciding with this period of diversification. The continents we observe today continued to be shaped into their current positions by the movement of tectonic plates after fragmentation from Laurasia and Gondwana.

Eocene Epoch

56 to 33.9 million years ago

The Eocene Epoch began with the Paleocene-Eocene Thermal Maximum (PETM), a period of intense warming that lasted nearly 170,000 to 200,000 years. It was a time of warm and wet climates with lush forests extending into higher latitudes, which gradually transitioned into a cooler, drier climate during the Late Eocene to Early Oligocene. The continents continued to break apart from Laurasia and Gondwana as Australia and Antarctica began to separate in the Early to Middle Eocene. India would eventually collide with Asia, initiating the uplift and formation of the Himalayan mountain range. These tectonic shifts dramatically impacted the transition of Earth's oceans and overall climate to cooler and drier conditions by the end of the Eocene Epoch. This cooling trend facilitated the expansion of the Antarctic ice sheet and the transition of Earth to an icehouse climate (Eldrett et al. 2009; Pound and Salzmann 2017). It also led to more prominent seasonal changes and the continued development of open woodlands with grasses reshaping these environments and their inhabitants. However, the greenhouse conditions of the Early Eocene, with generally warmer seas, contributed to the rapid radiation of marine mammals, while coastal environments became vibrant ecosystems with fruit-bearing flora. Invertebrates, reptiles, and sharks all filled ecological niches in the rich marine landscapes of the Early Eocene, which contributed to the diversification and interplay between coastal and inland ecosystems during this epoch. The later climatic cooling of the Eocene-Oligocene transition (EOT), which was the Cenozoic Era's most intense cooling event, marked a significant faunal turnover in the Eocene-Oligocene extinction event, particularly in Europe, known as the Grand Coupure. It is evident in the Eocene-Oligocene boundary both on land and in marine stratigraphy.

While the end of the Eocene marked a dramatic impact on the biodiversity on Earth for a brief time due to the extinction event, early true primates with adaptations for arborealism emerged during this epoch, including adapids and omomyids, which were small lemur-like creatures. These adaptations included stereoscopic vision for increased depth perception, along with grasping hands and feet, which aided in navigating woodlands. Fossils of these primates have been discovered in North America, Eurasia, and Africa (Smith, Rose, and Gingerich 2006). Adapids are considered the ancestors of modern strepsirrhines, such as lemurs and lorises, with extinct species like *Notharctus*, *Darwinius*, and *Adapis* resembling these modern primates. Omomyids are considered a potential stem group rather than direct ancestors to modern haplorhines, which include tarsiers, monkeys, and apes. Species of omomyids, such as *Teilhardina* and *Cantius*, exhibited features similar to those of modern tarsiers, including larger eyes and brains relative to body size. The Eocene was a pivotal time period for the early development of primate species, although the intense cooling in northern latitudes at the end of the epoch likely reorganized and confined the geographical reach of primates. However, this also set the stage for major primate groups to continue evolving in more limited regions.

Oligocene Epoch

33.9 to 23.03 million years ago

The Oligocene Epoch began with the Eocene-Oligocene Transition (EOT), where Earth experienced cooler and drier climates, ultimately leading to a growth in the ice caps and the glaciation of Antarctica. This lowered sea levels, exposing landmasses that had previously been underwater. It was the Cenozoic's largest cooling event and saw a significant global drop in carbon dioxide concentrations, switching Earth's climate from greenhouse to icehouse conditions (Hutchinson et al. 2021). While many species of marine life and terrestrial mammals experienced extinction due to this, the global expansion of grasslands during this time had a significant impact on the diversification of ecosystems.

As warm and temperate seasonal forests spread in many regions, the first early apes, known as hominoids, evolved from earlier lineages of primate ancestors. This was a crucial point in mammalian evolution, which led to the rise of great apes, or hominids, in the Miocene. Other terrestrial life became prominent during the Oligocene, including grazing animals, carnivores, and rodents, which shaped many ecosystems that early primates became a part of in the Miocene. This consequently influenced the evolutionary path of hominids.

Miocene Epoch

23.03 to 5.3 million years ago

The Miocene Epoch was generally considered to be much warmer than the preceding Oligocene, with major grasslands beginning to span the globe. The Middle Miocene Climatic Optimum (MMCO) was a period of warm and humid temperatures that was directly followed by cooling and ice-sheet expansion, particularly of Antarctica, during the Middle Miocene Climatic Transition (MMCT). The savanna hypothesis posits that the expansion of grasslands facilitated the development of bipedalism in primates due to adaptations with predation, visibility, and long-distance travel, giving rise to the first hominins around 6 to 8 million years ago.

This hypothesis is challenged by some early hominin fossils, such as *Ardipithicus*, that have been associated with tree or bush savanna habitats, suggesting that the rise of bipedalism in early hominins was not solely driven by the need to adapt to open grasslands (Cerling et al. 2010). The focus of bipedal evolution in early hominins has shifted to a more complex understanding of seasonal mosaic environments. These environments may have included both riparian forests and grasslands, where seasonal changes influenced the distribution of resources. A sister taxon to the lineage leading to modern humans also emerged in Eastern Africa during the Miocene, which included *Proconsul africanus*, *Proconsul gitongai*, *Proconsul major*, and *Proconsul meswae* between 23 and 14 million years ago. These basal (early) hominoids are significant in that their fossils offer crucial insights into the evolution of primate locomotion, particularly the forelimbs. They represent a grade of evolution close to the split of monkeys and apes, offering evidence into the ancestral characteristics of both groups.

The Late Miocene was a key time period of hominin evolution and the diversification of African apes. Tectonic activity in the East African Rift System (EARS) led to an uplift of the Eastern Arc Mountains, blocking moisture-laden wind from reaching west of the East African Rift Valley (EARV). As these winds moved inland from the Indian Ocean, they were forced to rise over the Eastern Arc Mountains causing air to cool and condense, known as orographic lift. This led to heavier rainfall on the eastern slopes of the mountain range, creating well-watered, fertile conditions in forested environments, which would have supported a wide range of biodiversity (Couvreur et al. 2020). In contrast, air that crossed the peaks of the range would have lost most of its moisture, leaving the land west of the slopes significantly drier, known as a rain-shadow effect.

The overall decrease in regional rainfall contributed to an expansion of grasslands and patchy woodlands at the floor of the rift valley. This was intensified by the atmosphere's reduced moisture-holding capacity following global isolation and expansion of the Antarctic ice sheet during the MMCT, which led to pronounced global cooling, seasonality, and aridity. These combined factors created an increasingly arid region in Africa during the Late Miocene, known as the African Arid Corridor, which stretches from the Horn of Africa to Namibia in the southwest. Throughout the Late Miocene and Early Pliocene, the expanding arid corridor acted as a barrier that fragmented the once continuous forest ecosystems of East and Central Africa (Bobe 2006). As a result, isolated highlands and coastal forests developed in the East, while forest patches were replaced by woodlands and grasslands in the rift valley, creating a diverse mosaic of seasonal habitats. This likely promoted the diversification and speciation of African apes and early hominins as they adapted to new ecological niches in this region.

Pliocene Epoch

5.3 to 2.58 million years ago

The Pliocene Epoch was a time of continued cooling and drying climates with the beginning of the Northern Hemisphere glaciation in the Eurasian Arctic and Northeast Asia, as well as the expansion of Antarctic ice sheets. This transition would aid in the establishment of Earth's modern seasonal climate pattern, with more pronounced glacial and interglacial periods facilitating the movement of megafauna across continents, including mammoths, which emerged in Africa in the Late Miocene-Early Pliocene. However, the Pliocene Epoch also experienced brief warming periods and fluctuating temperatures, particularly between 3 and 4 million years ago, known as the Mid-Piacenzian Warm Period (mPWP). Earth's climate was 2 to 3 °C warmer than today as a result of increased carbon dioxide levels in the atmosphere, providing insight into the effects of human emissions. This has raised concerns about potentially higher carbon dioxide levels in the atmosphere than at any time in the last 3.3 million years by the second half of the 2020s (de la Vega et al. 2020). The similarities in climate of the mPWP to modern-day conditions have become a valuable retrospective tool for understanding and predicting the future forecast of Earth's post-industrial climate.

Earth underwent significant geological changes throughout the Pliocene that would begin to reshape the structure of animal kingdoms in the North and South American continents. These changes were the result of ongoing geological processes set in motion long before the Pliocene. Between 40 and 38 million years ago, volcanic activity resulting from the subduction of the Pacific-Farallon tectonic plate beneath the Caribbean and South American plates created an underwater volcanic arc on the Caribbean boundary, referred to as the Panama Arc. This formation eventually collided with the South American tectonic plate and contributed to uplift in the seafloor and a shift in volcanic activity (Barat et al. 2014). Over millions of years, ocean currents brought accumulations of sediment, and the formation of the Panama Land Bridge finally closed off, connecting the ecosystems of both North and South America and facilitating the movement of animals between continents, known as the Great American Biotic Interchange (GABI). This migration of species over the Panama Land Bridge is estimated to have accelerated around 2.8 million years ago at the end of the Pliocene (O'Dea et al. 2016).

Hominins also became more diverse during this epoch, with species of both *Ardipithecus* and *Australopithecus* lineages emerging in the late Miocene and early Pliocene in Africa. The continued evolution of early hominin ancestors in the Pliocene becomes more defined towards the end of the epoch when the first evidence of *Homo* emerges in the fossil record between 2.8

and 2.75 million years ago. Though, as aforementioned in the section discussing the Neogene Period as a whole, more concrete evidence of *Homo* species, such as *Homo habilis*, is found to have emerged approximately 2.4 million years ago during the beginning of the Pleistocene Epoch. The Pliocene also marks the early stages of the Stone Age in Africa, where early hominins began shaping rudimentary tools to assist with subsistence activities. The exact time when these technologies were used by hominins is still part of ongoing research that is fragmented by the limited archaeological record, but our earliest evidence brings the use of stone tools by our ancestors back as early as 3.3 million years ago.

Pleistocene Epoch

2.58 million to 11,700 years ago

The Pleistocene Epoch is often associated with Earth's most recent ice age, the Pliocene-Quaternary glaciation, which began approximately 2.58 million years ago. Early hominins continued to diversify from australopithecine lineages during this time, with the first *Homo* species, *Homo habilis*, appearing in the fossil record around 2.4 million years ago. It is estimated that a significant population bottleneck lasted between 930,000 and 813,000 years ago, coinciding with longer glacial cycles of the Early Pleistocene and severe climate change. Studies using genomic sequencing found that this reduced the breeding population of human ancestors to approximately 1,280 breeding individuals (Hu et al. 2023; Muttoni and Kent 2024). This left a limited fossil record in both Eurasia and Africa between 950,000 and 650,000 years ago.

The Last Glacial Period of the ice age spanned from roughly 115,000 to 11,700 years ago. The peak of the Last Glacial Period, known as the Last Glacial Maximum (LGM), occurred between 26,500 and 19,000 years ago. During this time, the Bering Land Bridge connected Asia and North America. Lower sea levels caused by large amounts of water locked in glaciers exposed the landmass, which persisted between approximately 36,000 and 11,000 years ago. This facilitated the migration of humans to the Americas, with the earliest Paleoindian occupation generally considered to have been between 25,000 and 12,000 years ago. This would mark the beginning of the Lithic/Paleoindian period throughout North, Meso, and South America until Early Archaic stone tool traditions emerged around 8000 BCE. As Earth trended towards a warmer climate coinciding with the end of the Last Glacial Period, a brief period of interruption in warming occurred approximately 12,900 years ago, known as the Younger Dryas. This intense cooling period lasted until about 11,700 years ago during the transition to the current interglacial period of the Holocene. By the end of the LGM, megafauna of the Late Pliocene and Early Pleistocene, including woolly mammoths, mastodons, and sabre-tooth cats, became gradually extinct in different regions.

QUATERNARY PERIOD

Holocene Epoch

11,700 years ago to present

The Holocene Epoch is the current epoch characterized by the current interglacial period on Earth. It began roughly at the end of the Paleolithic (Old Stone Age) after the last major cold episode of the Last Glacial Period around 11,700 years ago. As Earth's climate stabilized, humans gradually developed sedentary lifestyles and transitioned out of the Stone Age. Isolated populations of the woolly mammoth, along with other megafauna, would also become extinct around 4,000 years ago. As humans continue to have a significant impact on the global climate and Earth's ecosystems, it has raised the question of classifying another period as a distinctive epoch: the Anthropocene. However, to this date, it has been rejected as a formal geological epoch by the International Commission on Stratigraphy.

The following sections of this book will take the reader back to the previously discussed Miocene Epoch, which will cover human evolution, and the geological processes in which we have uncovered evidence for the justification of each species' inclusion into the human family tree. It will also cover the processes that facilitated the migration of early hominins out of Africa, and later human ancestors who migrated to the Americas during the Last Glacial Maximum (LGM). Following this transcontinental migration, the Holocene began by marking a transition into warmer and relatively more stable climates than the Pleistocene. The global shift in ecosystems was crucial for establishing favorable environmental conditions for the development of agriculture and settled communities.

A rise in sea level correlated with glacial melt from warmer temperatures changed coastlines and created new landscapes. Humans gradually came to dominate the Earth after adapting from the Stone Age technologies of our ancestors. Eventually, these changes transformed our way of life and began the search for answers about our place in nature. Our journey started as one defined by means of survival that, over thousands of years, became one of self-discovery.

BEFORE US: PALEOANTHROPOLOGY

Just 6 million years ago, a single female ape had two daughters. One became the ancestor of all chimpanzees, the other is our own grandmother.

- *Yuval Noah Harari* [Sapiens]

OUR STORY BEGINS 6 TO 8 MILLION YEARS AGO during the Neogene Period, when the first hominin-like mammals emerged. This would mark the divergence of early hominins from the last common ancestor (LCA) of great apes, chimpanzees, and gorillas. Understanding the evolution of extinct human ancestors is vital to complete the picture of the Geologic Time Scale and its correlation to prehistoric archaeological time periods. A fundamental aspect of differentiating these two distinct ways of classifying time periods lies in some of the earliest evidence of the Stone Age. It is here that the impact of hominins and early humans left significant traces of past activity that drive the symmetry of geoscience and anthropological inquiry, which provide the foundations for archaeological investigations.

To better understand the deep roots in human ancestry, the reexamination of differing theoretical approaches in archaeology by incorporating insights from paleoanthropology has heavily influenced "the prehistoric turn" as a focus on our paleoanthropological past before the creation of written records. Humans and technology are constantly and reciprocally shaping each other, where neither exists as a fixed, independent entity. Instead, the relationship between humans and technology evolves and is defined in relation to one another. By investigating the earliest development of stone tool technologies and their symbiotic evolution with early hominin users, and later archaic *Homo sapiens*, we find that humans have always shared a mutually constitutive relationship with technology (Coté 2012). This challenges ideas about the reliance on and integration with technology being distinctly characteristic of the digital age. Moreover, "the prehistoric turn" underscores the body's entanglement with technological environments and material culture.

What follows in this section is a general reference for twenty-eight of our known hominin ancestors, along with some of the prevailing fossil evidence for their inclusion into our evolutionary ancestry. It includes sources on their paleoenvironments, site information, and absolute dating methods among peer-reviewed factual interpretations of paleoanthropological and geoarchaeological data, which have ultimately led to our contemporary understanding of the odyssey of humankind. While there are topics of debate regarding hominin evolution and its linkage to our ancestry, like the absence of a C/P3 honing complex in hominin dentition and morphological characteristics of bipedalism, the following species are generally accepted as part of the hominin clade. It is also important to understand the distinction between the terms *hominid* and *hominin*. Hominids encompass all great apes, including humans, gorillas, chimpanzees, and orangutans, and their immediate ancestors. Hominins refer to hominini of the subfamily Homininae, which includes modern and extinct humans and their direct ancestors. All hominins are hominids, but not all hominids are hominins. Every year, more evidence is discovered that enables paleoanthropologists and archaeologists to further decipher and reconstruct the hominin lineage. The reader is implored to investigate the sourced information for themselves for a deeper understanding of hominin evolution.

“Toumaï”

Sahelanthropus tchadensis

7.2 to 6.8 million years ago

The emergence of upright bipedal locomotion in early human ancestors is estimated to have developed during this time frame, falling within the geological Neogene Period. This comes from evidence of the oldest bones attributed to *Sahelanthropus tchadensis*, which were dated between 7.2 and 6.8 million years ago. The remains, including a cranium among other fragments, were discovered in 2001 at the Toros-Menalla site in the Djurab Desert of Chad (Brunet et al. 2002). The nickname “Toumaï,” meaning “hope of life” in the local Dazaga culture of Chad, was given to the fossils when they were discovered by a team of paleoanthropologists led by Michel Brunet. The name “Toumaï” bears significant cultural meaning to the Dazaga culture, as it is given to infants who face difficulties and low chances of survival when born before the dry season in Chad. This symbolized the fragile hope for life that the fossil discovery of *S. tchandensis* represents in tracing human lineage and understanding the origins of humankind. This discovery pushed the timeline of human evolution further than previously understood.

Initially, the fossils were estimated to be between 7 and 6 million years old. It was not until a 2008 study by Lebatard et al. that the age estimate of the fossil assemblage was refined using the 10Be/9Be ratio cosmogenic nuclide dating method. This method measures beryllium-10 and beryllium-9 isotopes to determine the exposure age of rocks and sedimentary deposits at Earth's surface or the duration of their burial. Beryllium-10 is a radioactive cosmogenic nuclide produced in the atmosphere and in minerals like quartz by the interaction of cosmic rays with oxygen and silicon atoms. It has a relatively long half-life of 1.387 million years. An exposure age can be determined by measuring the concentration of beryllium-10 isotopes in a sediment sample, which are produced when cosmic rays interact with the oxygen and silicon atoms present in rocks or soil. The stable isotope of beryllium, beryllium-9, which is naturally abundant in Earth's crust, acts as a normalization factor to account for variations in the delivery of beryllium-10 to the sample from the atmosphere or mineral matrix.

The study focused on the Toros-Menalla fossiliferous area, specifically TM 266, where *S. tchadensis* was first discovered, and other site locations like TM 254, Koro Toro (KT 12), Kollé (KL), and Kossom Bougoudi (KB 1), which helped establish the age of the fossils. The KT 12 site is also notable for the location of another hominin species fossil discovery, *Australopithecus bahrelghazali*. The dating method examined the ratio of authigenic (formed in place) cosmogenic beryllium isotopes in clay minerals of the lacustrine facies and its sections of anthracotheriid units above and below the ash tuff layer in TM 254, and the two levels in TM 266 bracketing the *S. tchandensis* cranium fossil. This yielded a calibrated age of 7.04 ± 0.18 Ma for *S. tchandensis* (Lebatard et al. 2008).

While *Sahelanthropus* was potentially one of the first upright walkers, hominins like *Ardipithecus*, emerging around 4.4 million years ago, are often considered the earliest hominin species for which we have more robust fossil evidence of transitional bipedalism. Fossil evidence of *S. tchandensis'* femur suggests that this species may have been somewhat bipedal and partially arboreal (Daver et al. 2022). This adaptation is often attributed to a time of continued drying and cooling climates when the first grasses appeared in the Miocene. The species exhibited a mix of derived and early human traits with a sloping face and prominent brow ridges reminiscent of apes. However, they also possessed early human-like characteristics, including a shorter middle face, small canine teeth, and a foramen magnum positioned underneath the skull, suggesting bipedal capabilities. The gradual expansion of grasslands during the Miocene and Pliocene is often attributed to playing a role in the evolution of upright posture in early hominins. The transition from primarily arboreal life in forests to open habitats of savannas likely favored an upright posture and bipedalism for visibility, predation, and long-distance travel according to the savanna hypothesis (Dart 1957). This paleoenvironment hypothesis is challenged by some early hominin fossils, like *Ardipithicus*, that have been associated with dense woodland habitats, suggesting that the rise of bipedalism in early hominins was not solely driven by the need to adapt to grasslands (White et al. 2010; Cerling et al. 2010; Su 2013; Dominguez-Rodrigo 2014). The focus of bipedal evolution in early hominins has shifted to a more complex understanding of seasonal mosaic environments. This idea proposes that early hominins evolved in environments that may have included both woodlands and grasslands, where seasonal changes influenced the distribution of resources.

"Millennium Man"

Orrorin tugenensis

6.2 to 5.8 million years ago

The first fossils of *Orrorin tugenensis*, discovered in 2000, date back to approximately 6.1 to 5.8 million years ago, with thirteen known fossils belonging to at least five individuals. The fossil assemblages were found at four separate sites of the Lukeino Formation in the Tugen Hills of Kenya by a team of paleoanthropologists led by Brigitte Senut and Martin Pickford (Senut et al. 2001; Sawada et al. 2002). Coinciding with the turn of the new millennium and their discovery date, the team nicknamed the species "Millennium Man," reflecting the first hominin discovery of the 21st century. The fossils at the Cheboit and Aragai sites date to 6.1 million years ago, while the fossils at the Kapsomin and Kapcheberek sites are from the upper levels of the formation, dating to around 5.8 to 5.7 million years ago (Senut et al. 2001). The Lukeino Formation was built from fluvial and shallow lacustrine sediments of the East African Rift Valley (EARV) throughout the Middle Miocene to recent. Lavas underlying and overlying the formation, as well as crystals from within the sediments, were dated using a radioisotopic age determination method known as potassium-argon (K-Ar) dating.

This revealed that fossil sediments at the formation were deposited between roughly 6.2 and 5.65 million years ago, while the crystals found within the deposit dated to 6.06 Ma (Pickford and Senut 2001). K-Ar dating measures the ratio of radioactive potassium-40 to its decay product, argon-40, particularly in volcanic rocks where argon gas accumulates as they cool and solidify. By measuring the amount of potassium-40 and argon-40 within a sample against the decay rate, the time since the rock layer last solidified can be determined. Researchers also employed magnetostratigraphy to date rock layers by measuring polarity reversals in Earth's relative magnetic field through subsequent depositions of rock and tephrostratigraphy to analyze the individual signature of each volcanic ash layer (tephra). Each tephra has a distinct fingerprint represented in its chemical composition, which can help establish a chronological framework for depositional events.

While likely having a degree of arboreal locomotion, fossil evidence of the proximal femur bone suggests this species of hominin to be capable of walking upright on two legs. The femur of *O. tugenensis* bears characteristics typical of bipedal locomotion, including the curvature of the femoral shaft and the presence of a calcar femorale. Initial reports indicated that the femoral bone was morphologically closer to that of modern humans than to australopiths (Pickford et al. 2002). However, this was later challenged after a 2008 study compared the femoral morphology to later hominins in order to reconsider phylogenetic placement. The study revealed that the femoral shape strongly resembled characteristics of hip structure in *Australopithecus* and *Paranthropus* while differing from apes and *Homo*. This highlighted the bipedal capabilities of *O. tugenensis*, but also indicated that this species was more closely related to australopithecines than to the genus *Homo* (Richmond and Jungers 2008). In 2013, another study reassessed the morphological characteristics of the proximal femur bone within a larger sample of apes and hominins. The results indicated that the femoral structure of both modern great apes and hominins evolved in different directions, with the femur of *O. tugenensis* appearing as an intermediate between Miocene apes and australopiths, indicating incipient bipedalism (Almécija et al. 2013). This femoral evolution was found to be represented in the primitive morphology of some earlier apes.

The paleoenvironment of *O. tugenensis* was likely characterized by a mosaic of patchy woodlands and shrublands, although there is no indication of closed-canopy forests in the Tugen Hills during the time the fossils were deposited. This is based on stable carbon isotope analysis of large herbivore tooth enamel carbonate from fossil teeth found in the Lukeino Formation, which revealed that early Kenyan hominins inhabited an ecosystem in this region where C3 and C4 vegetation was present, potentially resembling a wooded grassland (Roche et al. 2013). More open areas of these habitats would have allowed C4 plants to thrive in hotter and drier conditions in direct sunlight, while C3 plants likely remained in shaded areas or moist, cooler patches of woodland. The fluvial and shallow lacustrine deposits from which the Lukeino Formation was made may also indicate the presence of bodies of water nearby at the time of deposition, where C3 plants would have been more prevalent. The presence of these early hominins in woodland environments further substantiates evidence that bipedalism was not solely an adaptation to the spread of grasslands.

Ardipithecus kadabba

5.8 to 5.2 million years ago

One of the last known early hominins with fossil evidence in the Miocene, *Ardipithecus kadabba*, was first discovered in 1997 in the Middle Awash region of Ethiopia. The first fossils were found at the Adu Asa Formation, with the bulk of them within the Asa Koma Member of that formation. Fossil remains were also found at the nearby Kuseralee Member of the Sagantole Formation. The name 'kadabba' in the Afar language of Ethiopia means "oldest ancestor," even though earlier hominin species have been discovered to date. Initial fossil remains consisted of a right lower jaw fragment with a molar found by a team of paleoanthropologists led by Yohannes Haile-Selassie. While first associated as a subspecies of *Ardipithecus ramidus*, further investigations between 1997 and 2001 produced additional fossils with approximately eleven specimens recovered (Haile-Selassie 2001). They included jaw fragments and teeth, a hand, a toe, and arm bones, which represented at least five individuals, further solidifying this discovery as a distinct hominin species. In 2004, a publication examined the morphology and wear pattern of teeth associated with *Ar. kadabba*, indicating that it was in fact a distinct species from *Ar. ramidus* (Haile-Selassie, Suwa, and White 2004). *Ar. kadabba* fossils also display characteristics of bipedalism in the remains of the toe bone, which, unlike that of apes, bears a human-like shape. However, this species also displays characteristics of arboreal locomotion, including curved fingers and long arms. These features would have enabled this species to adapt to a complex landscape that included forests, wooded riparian grasslands, and floodplains along water margins to access a variety of food sources.

The fossils were dated to approximately 5.8 to 5.2 million years old through radiometric dating of volcanic ash layers associated with the sediments where the fossils were found. Researchers at the Berkeley Geochronology Center under director Paul Renne used the argon-argon (40Ar/39Ar) dating method, a refined version of potassium-argon dating that is based on the decay of radioactive potassium into radioactive argon. It is also suitable for dating volcanic rock or ash layers and uses neutron irradiation to convert potassium-39 to argon-39, allowing for a single measurement of argon isotopes to determine the age. This circumvents probable issues with potassium-argon dating due to it requiring separate measurements of potassium and argon, which can introduce potential uncertainties.

Due to the relatively sparse fossil record of *Ar. kadabba*, it is difficult to fully reconstruct its physical appearance and its exact evolutionary placement, although it is considered a likely ancestor of *Ar. ramidus*. The mixed characteristics of this hominin represent the capacity for arborealism in their wooded environment, while also exhibiting rudimentary features of potential facultative bipedalism. On the ground, these hominins may have walked upright but were still adapted to a tree-dwelling lifestyle. This contrasts with *Ar. ramidus*, which exhibits more derived features of bipedal capabilities, yet is not fully habitual or obligate in upright walking. However, the degree to which *Ar. kadabba* could utilize these bipedal capabilities still attests to the notion that these characteristics developed in a mosaic of wooded and grassland habitats.

“Ardi”

Ardipithecus ramidus

4.5 to 4.2 million years ago

Emerging in the early Pliocene, the first fossil fragments of *Ardipithecus ramidus* were discovered in 1992 in the Lower Aramis Member of the Sagantole Formation, in the Middle Awash area of Ethiopia. The initial remains were found by an international multidisciplinary research team led by Tim D. White and Yohannes Haile-Selassie, with subsequent investigations yielding more fragments of a complete skeleton by 1994. The findings were first published in 1994, with the species being described as *Australopithecus ramidus* (White, Suwa, and Asfaw 1994), and later changed to the *Ardipithecus* genus by 1995. Nicknamed “Ardi,” meaning “ground” in the Afar language, with the full name ramidus meaning “root,” symbolizing the crucial insight into the early stages of human evolution that this discovery contributed. Moreover, it provided paleoanthropologists with more fossil evidence of partial bipedalism among early hominins around 4.4 million years ago.

The partial skeleton of *Ar. ramidus* recovered between 1994 and 1997 provided further insights into early hominin evolution on the divergence of human and chimpanzee lineages.

In 2009, a collection of articles was compiled and published detailing the "Ardi" skeleton along with the 110 specimens representing 35 individuals found while examining the characteristics of their paleoenvironments (White et al. 2009). The skeletal structure of *Ar. ramidus* exhibits features indicating the species was capable of bipedal walking, alongside their partial arborealism in a largely woodland environment. However, geochemical analysis of the Aramis sediments supports a paleoenvironment with a moderate amount of grass cover that is described as a patchy woodland forest, drier and more open than a tropical rainforest, yet not fully representing a savanna or tropical rainforest (Boisserie 2010). The position of the foramen magnum on the skull also indicates a more upright posture compared to that of chimpanzees.

While fossil remains of *Ar. ramidus* have been found at Aramis dating to 4.4 million years ago, the Gona region in Ethiopia has also yielded fossils dated between roughly 4.5 and 4.2 million years ago. Researchers used argon-argon (40Ar/39Ar) radioisotope dating combined with analysis of volcanic ash layers, namely the Gaala Tuff Complex (G.A.T.C.) and the Daam Aatu Basaltic Tuff (D.A.B.T.), where the fossils were found bracketed between to determine the dates. They also found the paleoenvironment of this species to have been in humid cool woodlands with moderate grass cover (Woldegabrial et al. 2009). The site of Aramis, where *Ar. ramidus* was found, located in the Afar Desert, is roughly 74 kilometers from where

Australopithecus afarensis fossils, including the famous "Lucy," were discovered. Although these two species existed at vastly different times from each other, the proximity of the fossil sites attests to the long-term pattern of hominin presence in this region of Ethiopia.

Australopithecus anamensis

4.2 to 3.8 million years ago

This species of hominin is the earliest known under the genus *Australopithecus*, or "southern ape," in which the prefix is derived from the Latin word *australis*, meaning "southern," and the suffix derived from the Greek word *píthēkos*, meaning "ape." The first significant fossil assemblages were discovered in 1994 at the Kanapoi Formation, located in the Turkana Basin of Kenya. An initial discovery at this site in 1965 yielded an unidentified arm bone; however, it was not until the team led by paleoanthropologist Meave Leakey in 1994 uncovered substantial fossil evidence for a new hominin species at the formation. Remains of nine dental, cranial, and postcranial specimens were recovered from the Kanapoi site, and 12 specimens from the Allia Bay site in Kenya (M. G. Leakey et al. 1995). Investigations within other sites at the Kanapoi Formation and Allia Bay area from 1995 to 1997 provided potential evidence that *Australopithecus anamensis* may have been an ancestor to *Australopithecus afarensis* (M. G. Leakey et al. 1998). However, later fossil discoveries of a cranium dating to 3.8 million years ago at the Woranso-Mille area of Ethiopia expanded evidence of this species' geographic range. This provided insight into geographically isolated populations of this species, which may have followed a separate evolutionary trajectory from other populations. This challenged the idea of anagenesis between *Au. anamensis* and *Au. afarensis*, revealing significant differences between the two species and an overlap in existence for nearly 100,000 years (Haile-Selassie et al. 2019; Haile-Selassie 2021). These species likely coexisted during the same overlap in time rather than evolved directly one after the other. Unlike other species, this hominin was not given a traditional or specific nickname based on the circumstances of its discovery.

Using single-crystal argon-argon (40Ar/39Ar) dating on the volcanic ash tuff layers that bracketed the fossil strata, researchers found an estimated range of 4.17±0.03 and 4.07±0.02 Ma (M. G. Leakey et al. 1998). The fossilized tibia of this species shows evidence of stress from weight-bearing, an indication of habitual bipedalism (Ward, M. G. Leakey, and Walker 2001). The wide area of bone below the knee joint, called the bicondylar angle, indicates that *Au. anamensis* was capable of walking upright, as well as standing on one leg. These individuals also exhibited strong and narrow jaws with both side rows of their teeth arranged in parallel lines. Sedimentological, faunal, and isotopic data indicate the early Pliocene paleoenvironments around the Kanapoi Formation were dynamic landscapes of closed woodlands, shrubs, and grasslands near permanent water sources, with the potential to shift rapidly between fluvial and lacustrine conditions (Bobe et al. 2020). The ecological and environmental conditions may have played a role in the rise of adaptations for habitual bipedalism in this species and other hominins that followed.

"Flat-Faced Man"

Kenyanthropus platyops

3.5 to 3.2 million years ago

Emerging in the Middle Pliocene, *Kenyanthropus platyops* is significant in that it helped challenge the idea of a single, ancestral lineage of hominins, demonstrating that a greater diversity of bipedal hominins coexisted in Africa between 3 and 4 million years ago. While sometimes classified as an australopithecine due to shared primitive traits like small ear holes, *Kenyanthropus platyops* also possesses modern features like a distinctive flat middle face and smaller molars, hence the nickname "Flat-Faced Man." Some of these australopith traits are exhibited in its contemporary, *Australopithecus afarensis*, leading to ongoing debate about the taxonomic classification of *Kenyanthropus*.

Some researchers maintain that due to the species' combination of primitive and advanced traits, it warrants classification as its own distinct genus from *Au. afarensis* (Spoor M. G. Leakey, and L. N. Leakey 2010; Spoor, M. G. Leakey, and O'Higgins 2016). The first fossils were discovered in 1998 and 1999 at the Nachukui Formation in the West Turkana region of Kenya by Meave Leakey and her research assistant, Justus Erus (M. G. Leakey et al. 2001). Specifically, this site is known as the

Lomekwi 3 site, notable for the discovery of the oldest known stone tools dating to around 3.3 million years ago. This important discovery now marks the beginning of the Stone Age in the archaeological record.

The remains included a partial skull (KNM-WT 40000) and a left upper jaw (KNM-WT 38350). The skull was found below the Tulu Bor Tuff, which was radioisotopically dated to roughly 3.4 million years old, and above the Lokochot Tuff, dated to 3.57 million years old. The left upper jaw specimen was found above the Tulu Bor Tuff and dated to approximately 3.3 million years old. This species of hominin is known to have a smaller endocranial volume (350 to 500 cc) (inner lining of the skull) compared to other hominins of the same period, with adaptations that favored bipedalism but lacking substantial conclusive evidence for this type of locomotion. The fossil layers in which *K. platyops* was discovered suggest these individuals may have inhabited a mixed grassland and wooded environment.

"Lucy"

Australopithecus afarensis

3.9 to 2.9 million years ago

When many consider early hominin evolution and the subject of paleoanthropology, they often point to *Australopithecus afarensis*, the species of our Pliocene ancestors which has otherwise come to be known as "Lucy." *Au. afarensis* fossils have been found in Ethiopia, Kenya, and Tanzania between layers of volcanic ash tuff, known as tephra, with date ranges corroborating its existence in the Pliocene between 3.9 and 2.9 million years ago. The most famous fossil discovery of this species, given the nickname "Lucy," was one of the most complete early hominin skeletons recovered at around 40% completeness. Found in 1974 at the Hadar Formation in Ethiopia, Lucy's fossils date to roughly 3.2 million years old (Johanson, White, and Coppens 1978). The remains were discovered by a team of researchers led by paleoanthropologist Donald Johanson, along with then graduate student Tom Gray. The fossils challenged original theories about early hominin bipedalism, indicating it may have actually evolved much earlier than larger brains (Harcourt-Smith 2010). This idea is further solidified by the Laetoli Scans, a set of hominin footprints preserved in volcanic ash, which were discovered in 1976 at Laetoli in Tanzania. The discovery at the time, made by paleoanthropologist Mary Leakey and her team, was the oldest unequivocal evidence of bipedalism in early hominin lineage dating back to roughly 3.66 million years old (Day and Wickens 1980). The footprints are most likely attributed to *Au. afarensis* (White and Suwa 1987), and demonstrate that bipedalism evolved early in hominin evolution long before evidence of larger endocranial volume appears in the fossil record. More recent studies on the morphological characteristics and gait of the footprints left at Site A and Site G indicate two separate hominin taxa may have coexisted at Laetoli (McNutt et al. 2021). The discoveries surrounding Lucy have slowly helped paleoanthropologists reconstruct the timeline of bipedalism in early hominin species. They have also provided crucial insights into the nature of coexistence between hominin species that overlapped in time with each other throughout the same regions.

Considered to have lived around the same time as *Kenyanthropus platyops* in East Africa, fossil evidence suggests that both *K. platyops* and *Au. afarensis* may have used stone tools. The earliest evidence for their use by *Au. afarensis* was found in Dikika, Ethiopia, where bones bearing cut marks attributed to stone tools were uncovered, dated using argon-argon (40Ar/39Ar) isotopes back to roughly 3.39 million years old (McPherron et al. 2010). This also provided some of the earliest evidence of hominins consuming meat, marking an important evolutionary adaptation and pushing the known timeframe for these behaviors in human ancestors by nearly a million years. Other evidence of early hominin stone tool use was uncovered at the Lomekwi 3 site located on the western shore of Lake Turkana in northern Kenya (Harmand et al. 2015). The makers of the Lomekwian stone tools are ultimately unknown due to a lack of hominin fossils found at the site to definitively link a species to their creation. The most likely candidates are *Australopithecus* or *Kenyanthropus*, the two hominin species living in this region during the time. The evidence of stone tool use among these hominins, specifically the bones found in Dikika, indicates a gradual shift to the consumption of meat in early hominins.

The assemblage of cut bones at Dikika was found just several hundred meters away from another well-known *Au. afarensis* fossil discovery, and is dated to approximately 3.3 million years ago. These fossils were of a juvenile female *Au. afarensis* given the nickname "Selam," meaning "peace" in the Ethiopian Amharic language. The remains, including one of the most complete spinal columns ever found for an early hominin, were discovered in 2000 by paleoanthropologist Zeresenay Alemseged and his team at the Dikika site (Alemseged et al. 2006). Selam's assemblage also contained twelve pairs of ribs, a scapula, a hyoid bone, and clavicles, along with hand, foot, and limb bones, and a nearly complete skull. The morphological characteristics of the assemblage helped develop a more complete picture of *Au. afarensis* developmental growth and anatomy from early stages of life. They also provided insights into the combination of bipedal walking and arboreal locomotion in this species.

Australopithecus deyiremeda

3.5 to 3.3 million years ago

Australopithecus deyiremeda is another australopithecine that lived in Ethiopia during the same time as *Kenyanthropus platyops* and *Australopithecus afarensis*. The first fossils were discovered in 2011 in the Woranso-Mille area of the Afar Region in Ethiopia. The remains, including upper and lower jaw bones, were found in the Burtele and Waytaleyta sites of 3.5-3.3 million-year-old deposits within this region by paleoanthropologist Yohannes Haile-Selassie (Haile-Selassie et al. 2015). The holotype was formally named later in 2015 as *Au. deyiremeda*. The species name, *deyiremeda*, meaning "close relative" in the local Afar language of Ethiopia, was chosen because it is believed to be a close relative of other hominin species like *Au. afarensis*. This is due in part because both species coexisted in the Afar Region of Ethiopia around the same time, between 3.5 and 3.3 million years ago. However, *Au. deyiremeda* fossils exhibit a more robust jaw structure and thicker tooth enamel than *Au. afarensis*, warranting their distinct taxonomic classification in the hominin lineage.

While there is no direct evidence that *Australopithecus deyiremeda* used stone tools, there is strong circumstantial evidence that suggests they may have used them, given the discovery of stone tools at the Lomekwi 3 site and its contemporaneity with the existence of *Au. deyiremeda*. This species and its proximity to the other two hominin taxa associated with stone tool usage, *Au. afarensis* and *K. platyops*, also reinforce the idea that *Au. deyiremeda* may have been capable of developing them at the time. The coexistence of hominin species in the Woranso-Mille area of Ethiopia is attributed to the mosaic of Pliocene environments in the region, specifically a mix of grassland and woodland habitats, which contributed to specialized diets in these different ecological niches (Su and Haile-Selassie 2022). The paleoenvironment of *Au. deyiremeda* was likely a diverse seasonal mosaic environment composed of open grasslands and riparian forests, which contrasts with the Afar Region's current arid and open landscape.

"Abel"

Australopithecus bahrelghazali

3.6 to 3.5 million years ago

As one of the first hominins discovered in Chad until *Salhelanthropus tchadensis* was discovered in 2001, *Australopithecus bahrelghazali* marked a significant turning point in understanding the geographic distribution of australopithecines in Africa. This species of australopithecine also lived in roughly the same overlap in time as *Kenyanthropus platyops*, *Australopithecus afarensis*, and *Australopithecus deyiremeda*, although the remains discovered are the first from the genus to be found outside of southern and eastern Africa. *Au. bahrelghazali* was given the nickname "Abel" in memory of French geologist Abel Brillanceau.

Characteristics of this hominin include a relatively flat and vertical mandibular symphysis (chin area) compared to other australopithecines. The molar fossils of *Au. bahrelghazali* also exhibits two cusps (bicuspid), which further distinguishes it from other australopiths. While the discovery of these fossils in Central Africa indicates that australopithecines had a wider geographic range outside of southern and eastern Africa, the similarities shared with fossils of *Au. afarensis* has led some to classify it as a variant. However, the distinct morphological characteristics, particularly that of the wider mandibular symphyseal shape, has been pointed to as warranting the distinct taxonomic classification of *Au. bahrelghazali* (Guy et al. 2008).

The first fossils of *Au. bahrelghazali* were discovered near the town of Koro Toro in the Bahr el Ghazal region of Chad at the KT 12 site by paleontologist Michel Brunet and his team in 1995. This fossil assemblage included a jawbone (holotype KT12/H1) and an isolated premolar (paratype KT12/H2), which were used to determine the new taxa based on morphological characteristics. These fossils were dated using the 10Be/9Be ratio cosmogenic nuclide dating method, which yielded an age of 3.58±0.27 Ma (Lebatard et al. 2008). The fossils were found in poorly consolidated, peri-lacustrine clayey sandstone layers, which are associated with Lake Chad. This indicates that *Au. bahrelghazali* may have lived in a lakeside grassland paleoenvironment with sparse tree cover, relying on C4 grasses and sedges as subsistence (Macho 2015). As with other Pliocene hominins living near the time of the first stone tool development, questions have been raised about the possibility that *Au. bahrelghazali* had the capabilities to manufacture and use them for simple subsistence functions. While this species may have used stone tools, there is no direct evidence to support that they specifically used them in the archaeological record.

"Taung Child"

Australopithecus africanus

3.3 to 2.1 million years ago

Emerging in the Late Pliocene to Early Pleistocene in South Africa, *Australopithecus africanus* was one of the first early human fossils to be found in Africa. Its discovery by anthropologist Raymond Dart in 1924 challenged the idea that early humans had evolved in Eurasia. While initially Dart's discovery was met with resistance by proponents of Eurocentric theories about early human origins, later discoveries in Africa helped support the Out of Africa hypothesis gradually leading to a paradigm shift in understanding hominin evolution.

The first fossil, the "Taung Child," was one of the first early human fossil discoveries to draw attention to this region of Africa as a place of origin in the human family tree, dating approximately between 3.3 and 2.1 million years old (Dart 1925). The Taung Child's skeletal remains indicated that *Au. africanus* also walked upright, as the foramen magnum is positioned towards the front of the skull, diagnostic of bipedalism in hominins. The nickname, "Taung Child," was given to the fossilized skull because it was discovered at the Taung site, a limestone quarry near the town of Taung. The nickname reflects its place of discovery, but also the fact that this specimen was a juvenile.

The holotype specimen was discovered in a pink breccia deposit near a large, decalcified infill of reddish-brown sand. More specifically, in a tufa formation at Buxton Limeworks Quarry, which was created by calcium carbonate precipitating from dolomites. The skull was found within a calcrete and breccia matrix, creating a natural cement formed by water over time. Paleoanthropologists suspect the Taung Child was likely killed by a bird of prey, as puncture marks were found at the bottom of the three-year-old's eye sockets (Berger 2006).

When compared to earlier australopithecines like *Au. afarensis*, *Au. africanus* had a rounder cranium housing a larger brain and smaller teeth. No definitive evidence of stone tool use has been found that is directly associated with this species in its known fossil sites. Recent studies of the trabecular bone structure in *Au. africanus* hands indicate human-like hand posture and dexterity capable of using stone tools (Skinner et al. 2015a; Almécija et al. 2015; Skinner et al. 2015b). This has been the subject of ongoing debate; however, more definitive evidence of stone tool use is required for a conclusive answer.

Australopithecus garhi

2.6 to 2.5 million years ago

Emerging in the Early Pleistocene Epoch, it is generally believed that *Australopithecus garhi* lived around the same time as *Australopithecus africanus*, with recent studies suggesting *Au. garhi* may have been a side-branch of the human lineage and not a direct ancestor to *Homo*, as they do not share any derived traits (Strait and Grine 2004). The evolutionary line leading to modern humans is thought to have been a complex process involving multiple evolutionary pathways and potentially several species coexisting instead. However, it is still generally accepted based on the fossil record that there was a transition from the genus *Australopithecus* to an early *Homo* species between 3 and 2 million years ago, with *Homo* evolving from an ancestral australopith lineage.

The first fossils found, including a partial cranium, were discovered in the Hata Member of the Bouri Formation in the Middle Awash region of Ethiopia between 1996 and 1997. The remains were found by a research team led by paleontologist Berhane Asfaw and paleoanthropologist Tim White, with the partial cranium holotype specimen being discovered by paleoanthropologist Yohannes Haile-Selassie. The surrounding geological layers were dated using argon-argon (40Ar/39Ar) radioisotope dating, yielding an estimated age of 2.5 million years old (Asfaw et al. 1999).

Au. garhi exhibits larger teeth than other australopithecines, similar to *Paranthropus boisei*. Additionally, *Au. garhi* has been considered as a potential early user of stone tools, though evidence is debated and not entirely conclusive. Fossils of these individuals have been found near the same sites and geological layers as some of the earliest stone tools, along with animal bones bearing cut marks, suggesting this species may have also been among one of the first hominin species to butcher animals and consume meat (Asfaw et al. 1999). While both *Au. garhi* and *Homo habilis* are associated with the Oldowan tool industry, *H. habilis* is more definitively linked to the Oldowan tools. This is primarily due to the discovery of stone tools alongside *H. habilis* in sites like Olduvai Gorge, Tanzania. *Au. garhi* exhibits large premolars and molars, similar to *Paranthropus*, and its longer legs indicate the potential for bipedalism, although it possessed long arms capable of arborealism.

About 2.5 million years ago, the Middle Awash region in Ethiopia provided a landscape of scattered woodlands within savanna grasslands, which would have bordered a freshwater lake. *Au. garhi* likely inhabited a mosaic of woodland and grassland habitats like many other hominins, supporting diverse fauna such as grazing antelope and large predators. These environmental pressures could have led *Au. garhi* to rely on rudimentary stone tools for butchering and processing animal meat. While earlier evidence of meat consumption has been discovered associated with *Au. afarensis*, this is an important discovery related to the dietary adaptations of hominins over millions of years.

"Karabo & Issa"

Australopithecus sediba

1.97 to 1.78 million years ago

Northwest of Johannesburg, South Africa, lies the World Heritage Site known as the Cradle of Humankind. It is one of the most significant known paleoanthropological sites to date, yielding nearly 40% of all early hominin fossil evidence ever recovered. It is well-regarded for its particularly rich sites like the Rising Star cave system, Malapa, Kromdraai, Drimolen, Swartkrans, and Sterkfontein, containing fossil-rich deposits within dolomite formations. In 2008, a team led by paleoanthropologist Lee Berger, along with his then nine-year-old son, discovered the first remains of *Australopithecus sediba* in the Cradle's Malapa cave system. The first fossils were of a jawbone and collarbone belonging to a juvenile male (MH1), which was nicknamed "Karabo," meaning "answer" in the Tswana language. A nearly complete lower spine and teeth of an adult female (MH2) were also found at the same site, nicknamed "Issa," meaning "protector" in Swahili. The species name *sediba*, meaning "wellspring" or "fountain" in the Sesotho language, was selected by the discoverers to reflect the significance of this species as a source of information about our earliest evolution.

The remains were initially dated using uranium-lead (U-Pb) isotope dating by taking samples of the uranium-rich flowstone cap over the fossil layer. This method relies on the known decay rates of uranium-238 and uranium-235 into lead-206 and lead-207. By analyzing the ratio of uranium to lead in a sample with this established decay rate, the approximate age of when the mineral first formed and trapped any present lead can be determined. Researchers also combined this with paleomagnetic dating of sediments at the Malapa site, which compared their magnetic polarities with the established chronology of Earth's magnetic field reversals, revealing an estimated age range of between 1.95 and 1.78 million years ago (Berger et al. 2010). This period was also known as the Olduvai sub-Chron, where normal magnetic polarity occurred, meaning the Earth's magnetic field was in the same alignment we observe today. This is important as the Earth's magnetic field periodically reverses, switching its North and South poles, providing a reference point for understanding past climate changes. Moreover, the Olduvai sub-Chron is a major geological feature with the site of Olduvai Gorge in Tanzania, famous for its rich hominin fossil record. This chronostratigraphy has helped date sedimentary and rock layers found within the region of the gorge, as it is a key period for the emergence and dispersal of *Homo erectus*. Specifically, because this sub-Chron was a period of normal polarity during the Matuyama Reversed Polarity Epoch, which is characterized predominantly by a reversed magnetic field. A later study in 2011, combining uranium-lead dating and paleomagnetic sediment analysis of the flowstone matrix, yielded a refined date range of 1.977±0.002 Ma (R. Pickering et al. 2011a). This placed the *Au. sediba* fossil within an even shorter period of normal magnetic polarity, known as the pre-Olduvai Excursion. While Olduvai Gorge is located in East Africa and the Malapa site is in South Africa, this discovery demonstrates that *Au. sediba* fossils were deposited during a distinct and earlier period of normal magnetic polarity, which could have played a crucial role in the development of ecosystems during that time. The additional fossils of Issa recovered later in 2015 provided more complete insights about her lower back structure and lordosis, which is the natural curve of the cervical and lumbar regions in the spine. This has helped paleoanthropologists understand the transition towards obligate bipedalism and the emergence of human-like characteristics in australopithecines (Williams et al. 2021).

This species of australopith exhibits a mosaic of derived and early human features, with some traits closer to early *Homo* than *Au. africanus* or *Au. garhi*, further bridging the gap between earlier australopithecines and the genus *Homo*. Characteristics of the species include a small endocranial volume (420 to 450 cc), prominent brow ridges, and long molars relative to width, which are primitive traits. Some of the more defining derived traits include *Homo*-like tooth shape and growth patterns, increased manual dexterity, and a longer, more dexterous thumb. *Au. sediba* also possessed the ability of bipedalism, including a short, broad pelvis and a more vertical ilium. While *Au. sediba* is believed to have potentially used stone tools, there is no associated fossil evidence to support this idea. The paleoenvironment of this hominin is likely to have been a mosaic of grasslands and woodlands similar to other australopiths of this epoch. The region surrounding the Malapa site would have been cooler and much more humid, with a continued drying climate leading to the spread of savannas.

Paranthropus aethiopicus

2.7 to 2.3 million years ago

While *Paranthropus aethiopicus* is not considered a direct ancestor of *Homo*, it is more accurately described as a cousin or sister group to the *Homo* lineage. Both *Paranthropus* and *Homo* are thought to have descended from a common ancestor within the *Australopithecus* genus. *Paranthropus* is typically recognized as a side branch of its own genus, distinct from *Australopithecus* and *Kenyanthropus*, while *Homo* gave rise to modern humans. Even though *Paranthropus* is often grouped together with australopithecines, they are not always considered part of the same genus. Compared to *Au. africanus* and *Au. garhi*, two prominent species during this time, *P. aethiopicus* exhibits a strongly protruding face, large megadont teeth, and a well-defined sagittal crest on the top of the skull, which aided in the strength of its jaw. This indicates large crushing and chewing muscles. For these reasons, the *Paranthropus* genus is also sometimes referred to as the "robust australopithecines."

The first fossils of *P. aethiopicus* were described as *Paraustralopithecus* by paleontologists Camille Arambourg and Yves Coppens in 1967, based on a toothless mandible (Omo 18) found at the Shungura Formation in the Omo Valley of Ethiopia (Arambourg and Coppens 1967). It was not until 1985 that one of the more well-known fossil specimens, the "Black Skull" (KNM-WT 17000), was discovered in West Turkana, Kenya, by paleoanthropologists Richard Leakey and Alan Walker, which solidified the species into the *Paranthropus* genus (Walker et al. 1986). The skull was naturally stained black during fossilization due to the manganese-rich deposit it was found in. While initially identified as *Paranthropus boisei*, subsequent analysis revealed that KNM-WT 17000 possessed a mix of early human-like features resembling some australopiths, and derived features shared with *P. boisei and Paranthropus robustus.* This blend of features led to the identification of its own distinct taxa. The *Paranthropus* fossils have been dated using both potassium-argon (K-Ar) and argon-argon (40Ar/39Ar) radioisotope dating methods, yielding an age of 2.7 to 2.3 million years old, with the Black Skull dating to 2.5 million years old, placing this species in the Late Pliocene to Early Pleistocene.

Additional fossils of *P. aethiopicus* have also been discovered, including jawbones at the Koobi Fora Formation in Kenya, about 100 kilometers away from the Shungura Formation, along with an upper jaw and shin bone dating to 2.66 million years old at Laetoli in Tanzania. The Shungura and Koobi Fora formations are connected by their geology, as widespread volcanic ash tuff layers are found in both locations, which allow for direct lithostratigraphic correlation and dating. These tuff layers consist primarily of glass with phenocrysts of anorthoclase, hedenbergitic pyroxene, sodic amphibole, chevkinite, zircon, quartz, ilmenite, titanomagnetite, and more rarely, orthopyroxene and plagioclase (Martz and Brown 1981; Brown and Nash 2014). While there is no definitive fossil evidence linking *P. aethiopicus* to specific stone tool use or creation, evidence suggests they may have been capable of rudimentary stone technologies based on the findings of other *Paranthropus* species found near Oldowan stone tools at the Nyayanga site in western Kenya, dating to between 3.032 and 2.581 million years old (Plummer et al. 2023).

The paleoenvironment of *Paranthropus* was likely bushland to open woodland habitats with edaphic (waterlogged) grasslands, but not exclusively confined to riparian corridors. These semi-inundated habitats may have exhibited poor drainage and a low-moisture holding capacity, creating an ecosystem where it is more difficult for trees to survive even in the absence of major disturbances. However, research suggests these paleoenvironments were still trending toward drier and cooler conditions, leading to the spread of C4 vegetation. Isotope data from the enamel of *P. boisei* fossils discovered at Koobi Fora reveal that the C4 isoscapes this species relied on were dominated by grasses and sedge, which may have been used as a staple food source (Wood and Patterson 2020). This type of vegetation uses C4 carbon fixation photosynthetic pathways as an adaptation to hot and dry climates, allowing these plants to be more efficient in carbon dioxide uptake and reduce water loss compared to C3 plants.

"Nutcracker Man"

Paranthropus boisei

2.3 to 1.2 million years ago

One of the most significant and well-known paleoanthropological sites in the world is Olduvai Gorge, located within the Great Rift Valley of Tanzania. Notably, it has produced a wealth of evidence for early stone tool technologies and fossilized hominin remains, which include *Homo habilis, Homo erectus, Paranthropus boisei,* and *Homo sapiens*, which occupied the site much later. While the Cradle of Humankind World Heritage Site is located in South Africa, researchers have often attributed the name the "Cradle of Mankind" to Olduvai Gorge, located in East Africa, due to the remarkable discoveries of early hominin fossils and stone tools, which fundamentally rewrote our known history of stone tool technology and human evolution.

The first *P. boisei* fossils were discovered in 1955 by archaeologists and paleoanthropologists Louis and Mary Leakey at Olduvai Gorge. The significance of these fossils as a new species was at first not fully understood, as the assemblage lacked conclusive evidence for a new taxon. The initial fossils consisted of a deciduous canine and a large deciduous molar crown (OH

3) found in Lower Bed II. While the larger teeth suggested a potential megadont hominin present at the site, it was not until 1959 that Mary Leakey discovered the holotype specimen. The partial skull (OH 5), initially named *Zinjanthropus boisei*, confirmed their suspicions about the presence of a previously unknown hominin (L. S. Leakey 1959). The Leakeys recovered over 400 specimen fragments of the holotype skull, which became a landmark fossil, furthering our knowledge of human evolution.

An age of 1.75 million years old for the OH 5 cranium specimen was determined using potassium-argon (K-Ar) radioisotopic dating in the volcanic ash tuff layers found above and below the stratum where OH 5 was found. Stone tool assemblages were also discovered at the Olduvai Gorge site. Specifically, tools associated with the Oldowan tradition were found in the same stratigraphic layers as the fossil specimens. This stone tool technology typically dates back roughly 2.9 to 1.7 million years and is characterized by simple, unrefined stone tools consisting primarily of choppers, flakes, and scrapers used for processing plants and butchering animals. These stone tools were typically made from basalt and quartzite. While the Oldowan stone tool technology is often associated with *H. habilis*, also found at the Olduvai Gorge site in 1960, recent discoveries at the Nyayanga site on the Homa Peninsula in western Kenya, dating to between 3.032 and 2.581 million years old, included

Oldowan tradition stone tools alongside butchered hippopotamus bones and two *P. boisei* molars (Plummer et al. 2023). In the absence of *H. habilis* fossil assemblages at this site, it is generally thought *Paranthropus* could have been the creator of these Oldowan tools, but this still requires more conclusive evidence. While *P. boisei* anatomy suggests these individuals were capable of tool use, there is little definitive evidence that they specifically manufactured stone tools themselves throughout their existence in the Pleistocene Epoch. *P. boisei* fossils have also been discovered in other specific sites, including Peninj in Tanzania, Konso and Omo River Basin in Ethiopia, and Koobi Fora, Chesowanja, and West Turkana in Kenya. This includes a mandible (Peninj 1), which was found in 1964 at the Peninj site in Tanzania by Kamoya Kimeu, Glynn Isaac, and Richard Leakey (L. S. Leakey and M. D. Leakey 1964). Its discovery is significant as it represented one of the first examples of *P. boisei* to be found outside of Olduvai Gorge.

This hominin species received the notorious nickname "Nutcracker Man," which was attributed by the public over many years for its distinctive features. *P. boisei* had larger side rows of teeth than *Paranthropus robustus*, a flatter, larger endocranial volume than *Paranthropus aethiopicus*, and the thickest dental enamel of any known early human. Cheek teeth in this species are four times the size of a modern human's. *P. boisei* fossils also exhibit a well-defined sagittal crest in similarity to other hominins of this genus. The paleoenvironment of *P. boisei* was likely similar to that of *P. aethiopicus*, with a mosaic of bushland and open woodland habitats, and more closed riparian corridors near rivers and lakes. These environments are associated with C4 vegetation where grasses and sedges are readily available, as inundated conditions may have stunted tree cover, providing more direct sunlight and drier conditions for C4 plants to thrive. Isotope data from the enamel of *P. boisei* fossils discovered at Koobi Fora found that this species relied on a diet dominated by grasses and sedge, which may have been used as a staple food source (Wood and Patterson 2020). However, it is likely that *P. boisei* was not strictly confined to riparian environments, and their diet may have allowed them to explore a wide range of ecological niches.

"Robust Australopithecine"

Paranthropus robustus

2.0 to 1.2 million years ago

Another hominin of the *Paranthropus* genus emerging in the Pleistocene, *Paranthropus robustus* lived during the same time period as *Paranthropus boisei*, although they are not known to have lived simultaneously in exactly the same regions. Similar to *P. boisei*, however, *P. robustus* is known for its stockier build, large teeth, and defined muscles around the jaw. These characteristics are what earned this species and its associated genus relatives the nickname, "Robust Australopithecine." This species also exhibits a prominent sagittal crest on its cranium, similar to that of other megadont hominins.

The first fossils attributed to *P. robustus* were discovered in 1938 by Robert Broom in an outcrop of breccia deposit at the Kromdraai cave site in South Africa (Broom 1938). This cave system is located at the Cradle of Humankind World Heritage Site. The remains included the holotype specimen of a partial skull and jawbone fragment (TM 1517), teeth, a humerus, ulna, and toe bones. These fossils were dated to approximately 2 million years old using a combination of fauna biochronology in fossil strata and paleomagnetic analysis of sediment deposits within the site. Additional *P. robustus* fossils representing 99 individuals were found at the nearby Swartkrans Formation in Member 1, which includes the Lower Bank and Hanging Remnant deposits (T. Pickering et al. 2012). The age of the site was initially determined using uranium-lead (U-Pb) radiometric dating on sheets of calcium carbonate flowstone interbedded between clastic cave sediments. This provided bracketing ages of the fossil deposits in Member 1, which are underlain by these flowstone layers dated to 2.25±0.05 Ma and 2.25±0.08 Ma. These deposits are capped by layers dating to 1.8±0.01 Ma and 1.7±0.07 Ma, providing an approximate bracket age of 2.31 to 1.64 million years old, which was narrowed down using biostratigraphic data to between 1.9 and 1.8 million years old (R. Pickering et al. 2011b). Cosmogenic nuclide burial dating in further studies yielded a more precise age of 2.19±0.08 Ma and 1.80±0.09 Ma for Member 1 and 0.96±0.09 Ma for Member 3 at the Swartkrans Formation (Gibbon et al. 2014). This age range was found to be more precise than previous attempts at dating Member 3 using uranium-lead dating of tooth enamel (Balter et al. 2008) and electron spin resonance (ESR) data (Herries and Adams 2013); however, it remained consistent with the flowstone age found in Member 1.

This hominin, like others of the *Paranthropus* genus, is known to have been associated with bone tools, and there is evidence that they may have used unmodified stone and perhaps even manufactured some simple Oldowan stone tools. Fossil evidence of the *Paranthropus* genus alongside bone and stone tools at sites like Swartkrans, Nyayanga, and Drimolen raises the possibility of some level of sophisticated tool use. In an assemblage of 22 bone tools dated between 2 and 1.5 million years old, discovered alongside *Paranthropus* remains at the Drimolen site, 14 pieces of them exhibit the same wear patterns found on bone tools from the Sterkfontein and Swartkrans sites (Backwell and d'Errico 2008). The low number of stone tools found at Drimolen provides further supporting evidence that *P. robustus* was engaged in using these bone tools. More conclusive evidence that *P. robustus* was capable of cognitively using and creating bone tools came from the Cooper's D deposit, which is located between the fossil sites of Swartkrans and Kromdraai. The deposit is dated between 1.4 and 1 million years old, and has yielded seven *P. robustus* fossil remains and 50 stone artifacts alongside bone tools. In 2021, an analysis by Hanon et al. revealed similarities in morphological and use-wear patterns between a bone tool found at the Cooper's D deposit and bone tools previously identified at the Sterkfontein, Swartkrans, Kromdraai, and Drimolen sites (Hanon et al. 2021). This ultimately deepened our understanding of the various hominin species that may have been capable, physically and cognitively, of producing or using simple bone and stone tool technologies long before *Homo* did.

A new study, concerned with the importance of considering allometry (study of size and shape) when interpreting the paleodiversity of *P. robustus,* focused on the morphological characteristics in the temporal bone and semi-circular canals, analyzing size and shape variations (Braga et al. 2022). Researchers found that a specific *P. robustus* specimen from Unit P of the Kromdraai site is distinct due to a smaller temporal bone and semi-circular canals. The posterior semi-circular canals are also proportionally less developed, independent of age and sex. These findings highlight the potential size or shape variations when examining *P. robustus* fossil remains, particularly of the temporal bone and inner ear. This suggests localized adaptations to different environmental conditions within a relatively small geographical area, including open and well-watered mosaic grasslands with proximity to rocky outcrops and perennial river systems. The paleoenvironment in South Africa may not have been uniform, and *P. robustus* likely adapted to varying local conditions, potentially influencing their diet and behavior.

Homo rudolfensis

2.5 to 1.8 million years ago

Homo rudolfensis is considered an early *Homo* species, and while its fossils are dated earlier than some *Homo habilis* fossils, it is generally considered less well-defined and potentially more closely related to australopithecines. Compared to earlier australopithecines, *H. rudolfensis* exhibits a larger endocranial volume (750 to 825 cc), but its face and teeth remain relatively robust. This suggests a transitional stage in hominin evolution with a blend of features still reflecting australopithecine traits, while others point towards the *Homo* lineage. While no evidence of stone tool use or manufacturing has been directly found with *H. rudolfensis* fossils, it is considered highly probable that they used stone tools along with other early *Homo* species during the period of Oldowan tool use.

The first *H. rudolfensis* fossils were discovered in 1972 by Bernard Ngeneo, a member of Richard Leakey's team at the Koobi Fora Formation in Kenya. The holotype specimen (KNM-ER 1470) was a nearly complete skull, which challenged much of our understanding of *Homo* species paleodiversity (R. E. Leakey 1973). It was initially unattributed to a species of *Homo* until 1986 when anthropologist Valery P. Alekseyev proposed *Homo rudolfensis* in *The Origin of the Human Race*, for its place of origin in East Rudolf, Kenya (Alekseyev 1986). The dating of the volcanic ash layer where the holotype was found, known as the Kay Behrensmeyer Site (KBS) Tuff, was the subject of debate among paleoanthropologists for some time during the 1970s and 1980s, and is often regarded as the KBS Tuff Controversy.

It was named after geologist Kay Behrensmeyer, who discovered an assemblage of stone tools eroding out of the volcanic ash tuff layer in 1969. The KBS Tuff is a horizon of redeposited volcanic ash, meaning it contains a matrix of older and younger sediments as a result of fluvial, lacustrine, or colluvial depositional processes. Located in the East African Rift Valley (EARV), the holotype skull for this species was discovered below this volcanic ash tuff layer, which was initially found to have been contaminated with non-juvenile particulates that were inseparable from the sediment matrix. These original findings yielded inconsistent dates ranging from under 1 million to over 220 million years old. Additional argon-argon (40Ar/39Ar) dating of new sediment samples, where individual crystals could be separated, yielded an age of 2.61±0.26 million years old (Fitch and Miller 1970).

This age was challenged as it did not coincide with the established geochronology of the Omo Valley, making it difficult to correlate the Koobi Fora and Omo geological sequences. Later methods of biochronological analysis with pig molar remains from the site suggested an age closer to 2 million years ago

(Maglio 1972). At the British Museum of Natural History, paleontologist Alan Gentry used a similar technique with antelope remains, which also contradicted the 2.61 million-year-old date of the initial argon-argon dating. It was not until 1975, when geochemist Garniss Curtis and his student Thure E. Cerling at the Berkeley Geochronology Laboratory conducted argon-argon (40Ar/39Ar) and potassium-argon (K-Ar) dating in tandem with each other to discern that the original argon-argon dating on pumice deposits from the KBS Tuff horizon belonged to two separate sediments. They found that samples from Area 10 and Area 105 of the Koobi Fora Formation yielded an age of 1.60±0.05 Ma, and Area 131 with an age of 1.82±0.04 Ma (Curtis et al. 1975). The pig molar fossils from the 1972 study were also found to have been in the same stratigraphic layers as the KNM-ER 1470 holotype, which pointed towards a younger date for the KBS Tuff (White and Harris 1977).

Seeking to resolve the discrepancies, a study by Hurford et al. used fission-track dating of trace zircon separated from two pumice samples in Area 131, which yielded an age of 2.44±0.08 Ma for the eruption of the pumice (Hurford, Gleadow, and Naeser 1976). Fission-track dating analyzes the density, length, and number of damage trails left in the wake of spontaneous uranium-238 fission within crystalline structures. By knowing the uranium concentration within a sample, the age since it first cooled can be estimated based on the number of fission tracks present and their properties. This was more consistent with the original potassium-argon (K-Ar) and argon-argon (40Ar/39Ar) dating by Fitch and Miller, yielding an age spectrum of 2.61±0.26 Ma in Area 105, instead of the more recent potassium-argon date of 1.82±0.04 Ma for Area 131 in the 1975 publication. The study by Hurford et al. did not support the idea that there were pumice cobbles of different ages present within the KBS Tuff. Fitch et al. then corroborated this data with a recalculation of their original results, which yielded an estimated isochron age of 2.42 million years old based on an argon-argon (40Ar/39Ar) spectrum of 0.52±0.33 Ma and 2.6±0.3 Ma (Fitch, Hooker, and Miller 1976). They attributed the spread of the spectrum to a reheating of crystals after deposition. Further paleomagnetic investigations of the KBS Tuff horizon provided inconclusive results on the controversy surrounding these date spectrums (Brock and Isaac 1974; Hillhouse et al. 1977).

A trace element analysis of glass from pumice samples within the Koobi Fora Formation's KBS Tuff revealed a geological correlation between the H2 Tuff within the Shungura Formation, which dated to approximately 1.8 million years old (T. Cerling et al. 1979). The age spectrum of 1.60±0.05 Ma for areas 10 and 105 (Curtis et al. 1975) was later found to be in error, and a revised mean age of 1.8±0.1 Ma for the KBS Tuff was concluded from potassium-argon dating (Drake et al. 1980). This coincided with the contemporaneity between the KBS Tuff, the tuff layers of H2 and H4 in the Shungura Formation, and with Bed I at Olduvai Gorge, which establishes more conclusive age spectrum evidence for the KBS Tuff. Yet another fission track study was published following this in 1980 to eliminate any possible discrepancies, yielding an age of 1.87±0.04 Ma (Gleadow 1980).

Finally, in the early-mid 1980s, after the controversy had become quite convoluted, geologist Ian McDougall conducted independent potassium-argon (K-Ar) dating yielding an age of 1.89±0.01 Ma and argon-argon (40Ar/39Ar) dating yielding an age of 1.88±0.02 Ma (McDougall 1981; McDougall 1985). Additional magnetostratigraphic studies were also conducted on the Koobi Fora Formation, which refined the chronological framework for the hominin fossil-bearing sediments. This correlated upper magnetozones with parts of the Brunhes Normal Polarity Chron and Matuyama Reversed Polarity Chron, which are separated from magnetozones in the upper part of the Matuyama by a disconformity. The Olduvai Normal-Polarity subChron is also represented within the Matuyama; however, the lower part is missing due to an erosional disconformity (Hillhouse, Cerling, and Brown 1986). The increasing amount of geological evidence, along with the consistency of dates in recent studies, pointed ever more to the conclusion that the reheating of crystals following depositional processes at the site was actually unlikely.

McDougall later refined the argon-argon (40Ar/39Ar) dating geochronology in a 2006 publication discussing both silicic tuffaceous horizons present within the upper half of the

Koobi Fora Formation. McDougall and Brown found that pumice clasts within the KBS Tuff and Chari Tuff horizons, which bracketed the *H. rudolfensis* holotype, contained anorthoclase phenocrysts that are ideal for argon-argon (40Ar/39Ar) single-crystal dating. The feldspars found in pumice clasts in roughly 15 volcanic ash tuff layers (tephra) within the stratigraphic interval from the KBS Tuff to the Silbo Tuff, which is in the Chari Member, yielded more precise ages of the sedimentary sequences between the KBS (1.869±0.021 Ma), Chari (1.383±0.028 Ma), and Silbo Tuff (0.751±0.022 Ma) (McDougall and Brown 2006). These results provided a better framework for the upper geological sequences of the Omo-Turkana Basin as a whole, which in turn applies to a wider region in East Africa where these tuff horizons are also present.

This has brought us to the now generally accepted age of *H. rudolfensis* as living between 2.4 and 1.2 million years ago, with the holotype specimen from below the KBS Tuff to between 1.9 and 1.88 million years ago. The paleoenvironment of *H. rudolfensis* was likely a mosaic of open woodlands and savanna grasslands, adapting to the cooling and drying climate in East Africa. They likely coexisted with other hominins, including *H. habilis* and *Paranthropus boisei* where potentially different ecological niches were filled among each species. While the developing savanna environments were a major factor during this time period, *H. rudolfensis* was likely not confined to them, utilizing available forest resources when necessary.

"Handy Man"

Homo habilis

2.4 to 1.4 million years ago

Perhaps one of the first species of *Homo* with concrete evidence of stone tool use and manufacturing, earning this species the nickname "Handy Man," *Homo habilis* fossils have played a significant role in our understanding of early *Homo* species development. *H. habilis* exhibits smaller and more gracile traits, with a smaller face and teeth than *Homo rudolfensis*. They are also considered to have been a bipedal hominin, evident by traits such as a prominent arch in the foot with an adducted big toe. This species retained some ape-like features as well, including a longer arm span and shorter legs, suggesting they also had the capacity for arborealism in a paleoenvironment consisting of savanna grasslands with scattered trees and riparian avenues.

They are considered to be one of the earliest hominin species directly associated with the Oldowan stone tool industry, also known as Mode 1 tools. Oldowan stone tools are characterized by simple, flaked pebbles and worked cores that were used for various purposes, such as butchering and scraping animals, and processing plants. The first Oldowan stone tool discovery originally dated their creation back to approximately 1.8

to 1.7 million years old from Olduvai Gorge in Tanzania, attributed to *H. habilis* and potentially *Paranthropus boisei*. Recent investigations in the western Olduvai Basin uncovered another Oldowan assemblage dating to approximately 2 million years old (Stollhofen et al. 2021). This is about 150,000 years older than the first assemblages found in the eastern basin, pushing back the known timeline of hominin activity and tool use in Olduvai Gorge. However, Oldowan stone tools associated with the emergence of the genus *Homo*, found alongside animal bones bearing cut marks from the Gona region in Ethiopia, pushed the Oldowan technology date back earlier to between 2.6 and 2.5 million years old (Semaw et al. 1997; Semaw 2000; Semaw et al. 2003). Additional tools associated with *P. boisei* were found at the Nyayanga site in western Kenya, dating to between 3.032 and 2.581 million years old, which also included Oldowan tradition stone tools alongside butchered hippopotamus bones and two *P. boisei* molars (Plummer et al. 2023).

Early stone tool use has also been associated with australopithecines. Fossils of *Australopithecus garhi* have been found near the same sites and geological layers as some of the earliest stone tools, along with animal bones bearing cut marks, suggesting this species may have also been among one of the first hominin species to butcher animals and consume meat (Asfaw et al. 1999) Further evidence for their use by australopithecines was found in Dikika, Ethiopia, where bones bearing cut marks attributed to stone tool use by *Australopithecus afarensis* were uncovered, dated using argon-argon (40Ar/39Ar) isotopes back to roughly 3.39 million years old (McPherron et al. 2010). This provided even earlier evidence of evolutionary adaptations in hominins for consuming meat by nearly a million years. Furthermore, the discoveries of stone tools at the Lomekwi 3 site on the western shore of Lake Turkana in northern Kenya date to approximately 3.3 million years ago, and their creation is often attributed to *Kenyanthropus platyops* and *Au. afarensis* (Harmand et al. 2015). This pushed the timeline back for stone tool development among early hominins, predating that of the genus *Homo* by nearly 700,000 years. These discoveries fundamentally rewrote our understanding of when the Stone Age began.

The Oldowan stone tools were first identified in a 1931 publication by Louis Leakey, in which he noted "pre-Chellean" artifacts at Olduvai Gorge in the Great Rift Valley of Tanzania (L. S. Leakey 1931). It was not until 1936 that Louis Leakey formally described these early stone tools found at the gorge as Oldowan tools (L. S. Leakey 1936). The discovery of *P. boisei* fossils in 1959 at the Olduvai Gorge site prompted many to believe that the *Paranthropus* genus was the sole maker of these stone tools. This idea was challenged when the first *H. habilis* fossils were discovered in 1960 by paleoanthropologists Louis and Mary Leakey at Olduvai Gorge. The remains were initially found by their son, Jonathan Leakey, with a team composed of paleoanthropologist Philip Tobias and primatologist John Napier, who helped to analyze and reconstruct the remains. The holotype specimen (OH 7) included parts of a distorted lower jaw with teeth, an isolated maxillary molar, parietal bones (braincase), along with numerous finger, hand, and wrist bones (L. S. Leakey, Tobias, and Napier 1964). The age of the OH 7 was determined using potassium-argon (K-Ar) dating, yielding an approximate age of 1.75 million years old from the Lower Pleistocene layer of Bed I, where the fossils were found. These fossils were key to further distinguishing the *Homo* lineage from australopithecines. They were found alongside an assemblage of Oldowan stone tools, characterized by simple flakes, choppers, and cores created by hard-hammer percussion of basalt and quartzite.

More recent studies to revise the stratigraphic framework based on volcanic ash tuff geochemistry at Bed I of Olduvai Gorge used a combination of feldspar, augite, hornblende, and oxides found in glass and mineral compositions (McHenry 2012). This produced geochemical fingerprints for all ten major tuffs at Bed I with an estimated age of 2.03-1.79 Ma. The geochemical analysis also identified a more refined age of 1.84-1.79 Ma for *H. habilis* specimen OH 65, a well-preserved upper jaw and teeth, discovered in 1995 by anthropologist Robert Blumenschine within the western portion of Bed I (Blumenschine et al. 2003). This was based on the fact that OH 65 was recovered from a stratum between Tuff IC and the Ng'eju Tuff.

In 2014, Bernard Wood published "Human Evolution: 50 years after *Homo habilis*," which discusses the initial findings of the species and its implications for our understanding of human evolution to this day (Wood 2014). While *H. habilis* is still often attributed to the earliest stone tool creation and use, research over the years has developed a more nuanced picture of its stone tool use, suggesting the species may have had a more flexible diet utilizing these tools for a variety of processing needs, including both plants and animal meat. Others consider the morphology of *H. habilis* pertaining to its endocranial volume and skeletal features, which exhibit similarities between australopithecines. More fossils of early *Homo* species continue to be discovered, including a mandible and teeth (LD 350-1) from the Ledi-Geraru Research Project area in the Afar Region of Ethiopia, extending the record back to almost 2.8-2.75 million years ago (Villmoare et al. 2015). While not officially attributed to either *H. habilis* or *H. rudolfensis*, the early age of these fossils indicates a more temporally complex *Homo* lineage with some individuals exhibiting a broad range of australopith and early *Homo* traits (*Homo habilis sensu lato*), and some exhibiting unequivocal traits characteristic of the species (*Homo habilis sensu stricto*). Many of the more dramatic differences observed in *Homo erectus*, particularly a larger endocranial volume and taller stature, have positioned it as a more significant step in human evolution in the eyes of some researchers. However, the lineage of *Homo* is more likely a complex tapestry woven from multiple interactions and population dynamics across different regions and time frames that led to the development of early *Homo sapiens*.

Oldowan assemblages have also been discovered at Ledi-Geraru, dating to approximately 2.58 million years old (Braun et al. 2019). This discovery coincides with the variation in stone tool technologies around the same time it is observed in the Gona region and Nyayanga sites. Between 3 and 2 million years ago was a crucial time period for the development of stone tool technologies as the genus *Homo* and *Paranthropus* emerged, and their potential ancestor, *Au. afarensis*, who is also known to have used early forms of stone tools, disappeared around 2.9 million years ago. The variation in tool types indicates that early hominin technology was already diverse at this stage, rather than a single, uniform development. The presence of *Homo* fossils at 2.8 million years old in this site also coincides with the emergence of the genus *Homo* during this time and their influence on stone tool technologies. The proximity of these fossils to the stone tool assemblage supports a link between the emergence of the *Homo* genus and more complex early tool use. These tools also appear around the same time as hominin teeth begin to trend towards more *Homo*-like features, suggesting a connection between diet and technological necessity for survival.

Recent investigations at the Ledi-Geraru Research Project area revealed evidence of australopithecines and an early species of *Homo* living simultaneously in the region. More specifically, a new species of *Australopithecus* was identified through 13 fossilized teeth, though not formally attributed to official taxonomic nomenclature yet (Villmoare et al. 2025). This species was estimated to have lived between 2.8 and 2.6 million years ago and is not fully attributed to *Au. afarensis* due to distinct morphological characteristics of the teeth specimens, making the last known fossils of Lucy's species at roughly 2.95 million years ago. The identification of these fossils further supports that evolution was not linear and is a complex mosaic of branching pathways in our hominin ancestry. It also closes a complicated gap in australopithecine fossil evidence during this time period that was created by the presence of *Au. garhi*, also in the Afar Region of Ethiopia, between 2.6 and 2.5 million years ago. However, it marks an important discovery related to the evolutionary lineages of archaic *Homo* and australopithecines, revealing they in fact coexisted in the Afar Region simultaneously before 2.5 million years ago, rather than a scenario where *Au. afarensis* evolved directly into both early *Homo* and *Paranthropus.*

“Upright Man”

Homo erectus

2 million to 110,000 years ago

When *Homo erectus* emerged in the Pleistocene around 2 million years ago, a new stone tool technology appeared in Africa. These were purposefully shaped bifacial cutting tools characterized by large handaxes and refined choppers. This technology, the Acheulean (Mode 2), allowed *Homo erectus* to process food more efficiently, potentially contributing to their increased odds of survival in diverse environments where ecological roles were more niche. While known *Homo habilis* fossils remain exclusive to sub-Saharan Africa, *H. erectus* is the first hominin to have concrete evidence of migration out of Africa around 1.8 million years ago. Early African *H. erectus* fossils are the oldest early hominins to have possessed human-like body proportions, and these features are considered evolutionary adaptations to a life lived on the ground rather than a mosaic of arboreal and bipedal traits found in earlier hominins. *H. erectus* is the longest surviving species of early humans, existing for approximately 1.9 million years on Earth, significantly longer than even our own species.

The first *H. erectus* fossils, nicknamed "Java Man" were discovered in 1891 on the Indonesian island of Java by paleoanthropologist, Eugène Dubois, who initially classified the holotype specimens in an 1893 report as *Anthropopethicus erectus*, meaning "man-ape." The remains, including a molar (Trinil 1) and a skullcap (Trinil 2), were discovered in the Middle Pleistocene Kabuh Formation at the Trinil site in sediments exposed along the Solo River system. Dubois attributed them the name *Anthropopethicus* due to similar teeth fossils found in India at the time, which were also given this genus name to refer to ape-like ancestors. The following year after his initial discovery, Dubois found a femur bone (Trinil 3) at the site which led him to reexamine the initial findings. Later in 1894, Dubois reclassified the holotype to *Pithecanthropus erectus*, meaning "ape-human," after realizing the remains were more distinct from apes, representing a transition towards early hominins with a more erect posture (Dubois 1894). This term was originally coined by Ernst Haeckel to represent a hypothetical "missing link," which has since been superseded by a more nuanced understanding of the complexity in hominin evolution. Biologist Ernst Mayr later renamed the holotype specimens under the now accepted genus, *Homo erectus*, meaning "Upright Man" (Mayr 1950).

The first significant evidence that demonstrated *H. erectus* existed in other parts of the world was found between 1921 and 1923 at the Zhoukoudian cave site near Beijing, which has shown evidence of occupation between approximately 700,000 and 10,000 years ago (Sun 2024). The initial fossil remains consisted of two teeth discovered by geologist Johan Gunnar Andersson and paleontologist Otto Zdansky. These fossils were examined by paleoanthropologist Davidson Black, who identified a new species he called *Sinanthropus pekinensis*, often called "Peking Man" (Black 1927a; Black 1927b; Black 1929). Further excavations in 1929 led to the discovery of a nearly complete skullcap by paleontologist Pei Wenzhong. This prompted more excavations throughout the 1930s, yielding several skullcaps, stone tools that exhibited characteristics of both Oldowan and Acheulean technologies, and evidence of controlled fire use in layer 10, the earliest archaeological horizon at the site (Weiner et al. 1998). Soil micromorphology and Fourier transform infrared spectroscopy (FTIR) analysis of sediments at the site revealed that the alluvial deposition processes of fine silt-sized loess occurring in layers 10 through 3 included fine-grained organic matter (Goldberg et al. 2001). This indicated that the organic-rich deposits in layer 10 were actually a water-laid accumulation derived from outside of the site. Another study in 2022 used X-ray diffraction (XRD) analysis on 23 bones from layers 4 and 5, where the crystalline index revealed that at least 15 of the specimens were heated above 600°C (Huang et al. 2022). These findings have made the Zhoukoudian cave site one of the most productive *H. erectus* sites in the world, with approximately 45 individuals represented. The name *Sinanthropus pekinensis* was later reclassified by Ernst Mayr to *H. erectus*, at the same time that he renamed *Pithecanthropus* (Mayr 1950).

In 1971, paleoanthropologist Bernard Ngeneo discovered the remains of a lower jaw (KNM-ER 992) dated to approximately 1.5 million years old at the Koobi Fora Formation in East Turkana, Kenya. He also discovered a partial occipital bone (KNM-ER 2598) in 1974, which was below the KBS Tuff as well, and estimated to be 1.9 million years old. While KNM-ER 992 was initially classified as *Homo habilis*, it was later reclassified as *Homo ergaster* in 1975 after reexamination of the fossil, which was considered a regional variant of *Homo erectus* in Africa (Groves and Mazák 1975). Ngeneo subsequently discovered a cranium (KNM-ER 3733) in area 104 of the Koobi Fora Formation, which was also attributed to *H. ergaster* (R. E. Leakey 1976). It was also found in the same sediment layer as *Paranthropus boisei*, which has provided evidence for the coexistence between hominin species. The KNM-ER 3733 cranium has since been dated using magnetostratigraphy yielding an age spectrum of 1.63±0.15 Ma (Lepre and Kent 2015). This re-dating of KNM-ER 3733 created a roughly 300,000-year gap from 1.9 to 1.6 million years in the African fossil record of *H. erectus*. This gap places the KNM-ER 2598 specimen as the earliest African *H. erectus* fossil, and also older than the *H. erectus* specimens found at Java and those later found at Dmanisi.

In 1983, while excavating parts of a medieval city in Dmanisi, Georgia, archaeologists found fossilized animal bone assemblages, including rhinoceros teeth, which indicated a much older site than previously thought. Continued investigations in 1991 by anthropologist and archaeologist David Lordkipanidze revealed Oldowan stone tools and eventually, hominin remains of a lower jaw bone dating back to approximately 1.8 million years old (Gabunia and Vekua 1995). Further excavations revealed at least five individuals represented at the Dmanisi site in fossil assemblages, with one of the most complete Early Pleistocene hominin skulls (Skull 5) ever found outside of Africa, composed of a cranium (D4500) and mandible (D2600) (Lordkipanidze et al. 2013). While often attributed to *H. erectus*, Skull 5 exhibits a relatively small endocranial volume (546 to 775 cc) compared to other early *Homo* fossils. However, it also has a longer face and larger teeth, which are a blend of traits previously thought to have belonged to a different species. These uncommon characteristics in the Dmanisi hominin remains have led paleoanthropologists to question whether these variations are representative of a separate species (*Homo georgicus sensu stricto*) or are simply different traits within a single variable species like *H. erectus* (*H. erectus sensu lato*), and how to interpret these variations going forward. The discoveries at Dmanisi suggest that early *Homo* species were more mobile and adaptable than previously thought, potentially taking on different phylogeographic characteristics as they adapted to a variety of ecological niches after migrating out of Africa and experiencing isolation from other populations.

It was during this time of the Dmanisi discovery that additional *H. erectus* fossils were found in Africa, perhaps the most significant of them to date. In 1984, paleontologist Kamoya Kimeu, as part of a team led by Richard Leakey, discovered the remains of a young *H. erectus* male (KNM-WT 15000) dating between 1.6 and 1.5 million years old near Lake Turkana, Kenya, now known as the "Turkana Boy" (F. Brown et al. 1985). It is considered the most complete known early hominin skeleton ever recovered, with approximately 40% of the skeleton preserved. This has provided paleoanthropologists with a narrow window into the past to better understand *H. erectus* as a distant relative. It also provided more evidence to support the recognition of *H. ergaster* as distinct from *H. erectus*; however, this is an ongoing topic of debate, with *H. ergaster* often being attributed to exhibiting regional variances in Africa rather than an entirely distinct taxon. It is still important to recognize these distinctions and the ongoing discussion revolving around *H. ergaster* as its own taxa.

Compared to other early *Homo* species, *Homo erectus* had a taller and more vertical stature, increased endocranial volume (850 cc), and may have been the first early humans with direct evidence of controlled fire use. Evidence of burnt bones and plant material in cave sediments suggests possible controlled use of fire at additional sites like Wonderwerk cave in South Africa (Kaplan 2012). Micromorphology and Fourier transform infrared microspectroscopy (mFTIR) analysis of sediments from Wonderwerk provided evidence of burned bone and ashed plant remains in situ (found in place) dating to approximately 1 million years old (Berna et al. 2012). The use of fire and Acheulean stone tools by *H. erectus* is becoming increasingly probable given the recent studies, and they likely inhabited fluvial platforms between river valleys (river terraces) where this material was abundant. These adaptations, which marked some of the earliest beginnings of the Stone Age, demonstrate the understanding of stone properties, planning, and skill of *H. erectus*, indicating they had capable manual dexterity and advanced cognitive abilities of foresight. This versatility likely allowed them to also explore diverse environments while exploiting a wider range of resources, contributing to their overall survival and dispersal. Their paleoenvironments would have been different depending on the part of the world they occupied, including open woodlands, savanna grasslands, riparian corridors, steppe forestlands, and semi-arid xeric ecosystems with access to rocky outcrops.

"Pioneer Man"

Homo antecessor

1.2 million to 800,000 years ago

Homo antecessor likely emerged in Europe, potentially as a descendant of *Homo erectus sensu lato*, which includes African populations sometimes classified separately as *Homo ergaster*, or a more complex lineage of development following the large dispersal out of Africa around 2 to 1.8 million years ago by *H. erectus*. The name *antecessor*, meaning "pioneer," was attributed to this species due to its status as one of the earliest known hominin species in Europe, earning it the nickname "Pioneer Man." It is likely that *H. antecessor* contributed to the lineages of *Homo heidelbergensis* and Neanderthals. More recent molecular studies analyzing ancient proteins found on *H. antecessor* teeth suggest that it is not a direct ancestor of modern humans, but rather a closely related sister group to the common ancestor of modern humans, Neanderthals, and Denisovans (Welker et al. 2020). *H. antecessor* is characterized by a combination of cranial, mandibular, and dental traits, a somewhat larger endocranial volume (1,000 cc), with more modern midface features than those of *H. erectus* (Bermúdez de Castro et al. 1997).

The first fossils were discovered between 1994 and 1996 in the Atapuerca Mountains of Northern Spain at the Gran Dolina cave site by a team led by archaeologists and anthropologists Eudald Carbonell, Juan Luis Arsuaga, and José María Bermúdez de Castro. Specifically, the remains were found in stratigraphic unit TD6, which is situated below the Matuyama-Brunhes reversed polarity boundary, indicating the fossils were between 857,000 and 780,000 years old (Bermúdez de Castro et al. 1999). Stratum TD6 is also known as the "Aurora Stratum" after the Roman goddess of dawn, Aurora, signifying the dawn of the earliest evidence for human occupation in Europe. It contained a rich assemblage of butchered mammalian bones and other fauna, including those of other hominins exhibiting bone exploitation, bone breakage, peeling (a type of fracture similar to bending a fresh twig between the hands), along with percussion and chop marks, which has raised particularly horrifying questions in zooarchaeology concerning the first evidence of survival cannibalism over dietary needs (Fernández-Jalvo et al. 1999). These hominin remains found as part of a butchered faunal assemblage were associated with Mode 1 (Oldowan technology) stone tools in the Aurora Stratum, with nearly 30% of them exhibiting alterations, cut marks, or tool-induced damage (Díez et al. 1999). These tools included small flakes, large core-choppers, denticulate tools, and scrapers. The stone tool manufacturing associated with *H. antecessor* bears some similarities with the more complex Acheulean industry, but is still referred to as

Mode 1 technology. The purposes of early human cannibalism pertaining to the nature of human behavior have come into question by zooarchaeologists, and may indicate a desperate attempt for survival when resources were otherwise scarce.

Follow-up studies on the Gran Dolina cave site included sedimentological analysis of unit TD6, which classified its stratigraphy into three sub-units with 13 layers, containing nine defined sedimentary facies. The remains of *H. antecessor* were found in three of the nine facies, which are characterized as: debris flow facies, channel facies, and floodplain facies. These facies showed three different kinds of fossil distribution, including a group of scattered fossils, layers of fossils in fluvial facies, and a layer of fossils incorporated into fluvial and gravity flow facies. This study presented evidence that these facies were allochthonous (having originated from a distance), coming from an adjacent slope above the cave through fluvial processes. (Campaña et al. 2016). This could indicate why such a large faunal assemblage was found within the site; however, it does not qualitatively justify any lack of evidence for cannibalism at the site, regardless of depositional processes inside or outside of the cave, as indicated by the hominin remains bearing extensive bone exploitation.

In 2018, a study using electron spin resonance (ESR) dating on a tooth (ATD6-92) from the TD6 unit found an estimated range of 949,000 to 624,000 years ago. This was combined with magnetostratigraphic data from the unit to constrain the time frame produced from initial ESR chronology, resulting in a more refined age spectrum between 949,000 and 772,000 years ago for *H. antecessor* (Duval et al. 2018). The study supported genetic evidence for the antiquity of *H. antecessor* lineages predating the estimated time of divergence between modern and archaic human lineages.

Additional fossils have been found at the nearby Sima del Elefante site, namely a mandibular fragment (ATE9-1) in 2007, and an assemblage including a jawbone and fingerbone in 2008 (Bermúdez de Castro et al. 2011). In 2013, a set of nearly 50 footprints was found in Happisburgh, England, dating between 1 million and 780,000 years old. This extends the record of hominin occupation in Northern Europe by nearly 350,000 years. They are largely attributed to *H. antecessor* based on stone tool assemblages near the site, although no evidence of hominin fossils has been associated with the footprints (Ashton et al. 2014). The most recent discovery was in 2022, where archaeology graduate student Edgar Téllez discovered a significant partial face fossil (ATE7-1), nicknamed "Pink," which dated to roughly between 1.4 and 1.1 million years old, representing the oldest known human face fossil in Western Europe. The fossil of this hominin has been attributed as *Homo affinis* (relative to) *erectus* on the basis of this large gap in time and key morphological differences from those found in *H. antecessor* fossils at the

Gran Dolina site, which, as aforementioned, are dated to between 949,000 and 772,000 years ago (Duval et al. 2018). ATE7-1 exhibits a more primitive midface for the *Homo* clade, lacking modern features distinct in the Gran Dolina site remains (Huguet et al. 2025). Recent luminescence chronology studies have shown the upper Middle Pleistocene units (TE18 and TE19) contain Acheulean (Mode 2) and transitional lithic assemblages (Mode 3) alongside large and small mammal fossils which has aided in tracing human occupation patterns around the site, carnivore use of the cave, and the development of Lower Paleolithic stone tool technologies (Demuro 2022). While often considered to have been a potentially early user of fire, both sites of Gran Dolina and Sima del Elefante lack evidence to support that *H. antecessor* harnessed this capability or routinely used fire. The paleoenvironment of this hominin throughout Western Europe during this time was likely a warm, humid climate in a patchy landscape, which would have been primarily composed of humid meadows and riparian woodlands (Blain et al. 2010).

“Heidelberg Man”

Homo heidelbergensis

700,000 to 200,000 years ago

Although *Homo heidelbergensis* was formerly considered to have potentially been a last common ancestor of Neanderthals and *Homo sapiens*, the evolutionary lineage linked to modern humans is still part of continued research efforts in a culmination of genetic analysis and paleoanthropology (Mounier et al. 2009). Archaic *H. sapiens* instead most likely developed into anatomically modern humans as a result of interbreeding and genetic exchange among hominin populations in Africa between 200,000 and 60,000 years ago. These populations of archaic *Homo sapiens* then left Africa sometime between 125,000 and 60,000 years ago, known as the Out of Africa hypothesis (Ko 2016). There is an ongoing debate surrounding the classification of *H. heidelbergensis* as a distinct taxon based on their variable combination of plesiomorphous and synapomorphous traits, potentially distinguishing them from the Middle Pleistocene populations in Western Europe with derived Neanderthal morphology (Roksandic, Radović, and Lindal 2018). The taxonomic status of *H. heidelbergensis* has become more focused on Eastern Europe, as the new classification moves many African and some Asian fossils into *Homo bodoensis sp. nov*, which may encompass a broader group of hominins. This includes those previously classified as *Homo helmei* and *Homo rhodesiensis*, which are no longer definitively recognized as distinct species. This is aimed at creating a more unified classification of Middle Pleistocene hominins, particularly those found in Africa. It is worth noting that the 260,000-year-old "Florisbad Skull," attributed to *H. helmei* since its original discovery in South Africa in 1932, has now been associated with an early form of *H. sapiens* or a transitional form between *H. heidelbergensis* (*sensu lato*) and *H. sapiens* that existed in the Middle Paleolithic (Bruner and Lombard 2020).

H. heidelbergensis exhibits a larger endocranial volume than *H. erectus* (1100 to 1400 cc), and exhibits more robusticity in its skeletal structure that may represent transitional traits towards Neanderthals. Interestingly, *H. heidelbergensis* is also the species most likely responsible for the first projectile point spears, hafted to a wooden shaft. By this time, stone tool technologies were starting to trend towards more complex Middle Paleolithic industries, like the Fauresmith industry found in South Africa, which dates to approximately 500,000 years ago. It is some of the earliest evidence of blade technology in stone tools. *H. heidelbergensis* stone tool use is diagnostic of the Acheulean industry; however, they are known for later developing a more complex and conservative flaking method to produce finer tools known as the Levallois technique.

The first holotype fossil attributed to *H. heidelbergensis* was a mandible (Mauer 1) discovered in 1907 by Daniel Hartmann. It was found at the Mauer sand pit near Heidelberg, Germany, earning the species the nickname "Heidelberg Man." These sands consist of fluvial deposits of the Neckar River, which are overlain by glacial period loess deposits and interglacial period paleosols, indicating an early Middle Pleistocene age for *H. heidelbergensis* (Wagner et al. 2011). Subsequent fossil discoveries still attributed to the *H. heidelbergensis* species include a nearly complete skull with a face (Arago 21) from Arago cave in Tautavel, France, and a tibia (Boxgrove 1) from Sussex, England. Arago 21 was discovered in 1971 by Henry and Marie-Antoinette de Lumley, and is estimated to be around 450,000 years old. Boxgrove 1, otherwise known as part of the "Boxgrove Man" assemblage, which also contained two teeth found in subsequent investigations, was discovered in 1993 by archaeologist Mark Roberts. In a cross-sectional analysis of the tibia, researchers found that it exhibited a high level of robusticity within the range encompassed by Late Pliocene to Late Pleistocene archaic *Homo* combined with adaptations to colder climates, potentially associated with late interglacial cool temperatures of the region (Trinkaus et al. 1999). At the Acheulean site of Gesher Benot Ya'aqov in Israel, small burned flint fragments suggest burning occurred in specific spots, which has become an indicator for the controlled use of fire by early humans at the site 790,000 years ago (Goren-Inbar et al. 2004). This could have also been attributed to *H. heidelbergensis*, given their proximity in Eastern Europe.

Homo bodoensis

774,000 to 129,000 years ago

Recent reclassifications of many Middle Pleistocene hominin fossils from Africa and some from Southeast Europe, which were formerly categorized under the previously poorly defined *H. heidelbergensis (sensu lato* and *sensu stricto*) and *Homo rhodesiensis*, suggest they should now be considered *Homo bodoensis*. This new classification suggests *H. bodoensis* might be a direct ancestor to *Homo sapiens* and separated from a branch of lineage that later gave rise to Neanderthals and Denisovans. The reclassification was proposed by paleoanthropologist Miranda Roksandic at the University of Winnipeg in 2021, where she first made the case for *Homo bodoensis sp. nov*, indicating that both *H. heidelbergensis* (*sensu lato* and *sensu stricto*) and *H. rhodesiensis* should be abandoned as they do not represent the full range of hominin variability throughout the Middle Pleistocene in Africa, also known as the Chibanian Age (774,000-129,000 years ago). Roksandic et al. proposed that *H. bodoensis sp. nov* would represent an early Middle Pleistocene ancestor of the *Homo sapiens* lineage. This pan-African distribution would also account for fossils found in the eastern Mediterranean, including Southeast Europe and the Levant. They also proposed that many of the hominin fossils discovered in Western Europe that are currently assigned to the *H. heidelbergensis* taxa be reclassified as *Homo neanderthalensis* to reflect the transitional derived Neanderthal traits in Middle Pleistocene populations throughout the region (Roksandic et al. 2021; Harvati and Reyes-Centeno 2022). This notion of reclassification was subsequently rejected in a 2022 publication by paleoanthropologists Eric Delson and Chris Stringer, in which they cite Roksandic et al.'s argument as not meeting criteria for the International Code of Zoological Nomenclature (Delson and Stringer 2022). Roksandic et al. later followed up with a response defending their proposal to discontinue the use of *H. heidelbergensis* due to its problematic nomenclature and inconsistent use. They also argued for abandoning *H. rhodesiensis* due to its cultural insensitivity and highlighted the need for the requirements governing taxonomic nomenclature to be reviewed, given these complex subjects in paleoanthropology (Roksandic et al. 2022).

This ultimately coalesced from the convoluted state of Middle Pleistocene hominin taxonomy, commonly referred to as the "Muddle in the Middle," which is a key part in deciphering the lineage of early *Homo* species leading to *sapiens*. The specimen used by Roksandic et al. to identify this reclassification was a partial hominin cranium discovered in 1976 by Alemayhew Asfaw and Charles Smart, found in the Afar Depression at Bodo D'ar, Ethiopia (Kalb et al. 1980). The initial discovery, which was originally classified as *H. heidelbergensis*, included a lower face and part of the braincase estimated to be about 600,000 years old. This was followed by the upper facial portion discovered by Paul Whitehead and Craig Wood just weeks later. Interestingly, the fossils were found with an assemblage of Acheulean stone tools and faunal remains (Rightmire 1996), along with cut marks from stone tools on the holotype specimen's face and eye sockets, which some researchers point to as evidence of intentional defleshing.

The other notable taxon of *Homo* called into question by Roksandic et al. is *H. rhodesiensis*, which was first discovered in 1921 by Tom Zwiglaar. An intact skull (Kabwe 1) was found in a lead and zinc mine located in Kabwe, Zambia, which was part of Northern Rhodesia, earning the holotype the nickname "Rhodesian Man." The fossil was officially assigned to the taxon *H. rhodesiensis* in 1921 by paleontologist Arthur Smith Woodward. This specimen, along with the partial cranium identified at Bodo D'ar, came into question due to its variable morphological features, which made it distinct from those classified as *H. heidelbergensis* in Western Europe. *H. bodoensis* is not yet widely accepted as its own taxon based on the holotype reclassification by Roksandic et al., also due in part to the potential for its official classification to oversimplify human evolution. However, if officially recognized, it will represent Middle Pleistocene variations in hominin populations throughout Africa, the eastern Mediterranean, and Southeast Europe.

“Neo”

Homo naledi

335,000 to 236,000 years ago

Located at the Cradle of Humankind in South Africa is the Dinaledi Chamber of the Rising Star cave system, where over 1,550 specimens representing 15 individuals of *Homo naledi* were discovered in 2013. The team led by paleoanthropologists Lee Berger and Pedro Boshoff involved a reconnaissance mission by cave spelunkers Rick Hunter and Steve Tucker, reporting potential fossil finds to Berger. After squeezing through a narrow passage, the two discovered the previously unknown Dinaledi Chamber alongside hominin fossils scattered across the floor of the chamber. This chamber lies within the Monte Christo Formation, which is composed of thick stromatolitic (carbonate rock formed by cyanobacteria) dolomite horizons capped by a chert unit. The depositional age of the remains was discerned using a combination of uranium-series coupled electron spin resonance (US-ESR) dating on three teeth from the site and uranium-thorium (U-Th) dating of the oldest flowstone layer overlying the fossils. This revealed a depositional age spectrum of 335,000 to 236,000 years ago for the fossils (Dirks et al. 2017). Further exploration of the cave in 2017 revealed the nearby Lesedi Chamber, in which an additional 131 *H. naledi* specimens were recovered (Hawks et al. 2017; de Ruiter et al. 2019). These included a remarkably near-complete skeleton and cranium (LES1) nicknamed "Neo," meaning "gift" in the Sesotho language. The fossils of *H. naledi* revealed that this hominin represented some traits observed in australopiths, like smaller endocranial volume, but still exhibited similarities in cranial morphology with *Homo erectus*, *Homo habilis*, or *Homo rudolfensis*. The hand and wrist structure of *H. naledi* demonstrates human-like manipulatory adaptations, which are contrasted by australopith-like or *H. habilis* traits in the trunk, shoulder, pelvis, and proximal femur (Berger et al. 2015). Despite this, *H. naledi* still displays human-like feet and lower limbs, potentially representing the atrophy of some australopith traits in early *Homo*.

Although the anatomical capabilities of *H. naledi* are consistent with other early *Homo* stone tool makers, no direct evidence of stone tool use or manufacturing has been uncovered alongside their fossils. While there have been remnants of charred wood and bone recovered from the site where *H. naledi* was found, some researchers have raised questions about the dating of the burned material and the interpretation of the findings. Even though *H. naledi* could potentially be an early user of fire, there is no conclusive evidence to suggest they practiced controlled uses of fire. The large assemblages of fossils discovered associated with their deposition deep within the cave system, however, were presented as evidence to indicate deliberate body disposal in a single location by *H. naledi* (Dirks et al. 2015; Berger et al. 2023). This initial claim that *H. naledi* may have intentionally buried the dead was met with criticism, as the remains were found in a cave that experiences complex geological processes. This could have resulted in the accumulation and disarticulation of bones rather than a deliberate burial. In 2024, a publication argued there was insufficient sedimentological evidence to suggest that *H. naledi* engaged in mortuary practices or deliberate burials (Foecke et al. 2024). While the critique does not definitively rule out the potential for burial by *H. naledi*, it highlighted the shortcomings in the methodology and geochemical analysis of the initial claim.

“Neanderthals”

Homo neanderthalensis

400,000 to 40,000 years ago

Emerging in the Middle to Late Pleistocene, *Homo neanderthalensis*, otherwise known as Neanderthals, inhabited a wide geographic area across Eurasia between approximately 400,000 and 40,000 years ago. They are known to have lived alongside other hominins, including *Homo sapiens*, for a period of time. This overlap occurred in Europe and Asia, and genomic analysis has indicated there is evidence of interbreeding between Neanderthals and early modern humans between approximately 50,500 and 47,000 years ago, evident in some modern human populations still carrying Neanderthal DNA today (Sümer et al. 2024). Neanderthals are thought to have evolved from a branch of *Homo heidelbergensis* (*Homo bodoensis sp. nov*) that migrated to Europe and Asia from Africa, later diverging with a lineage leading towards Neanderthals in Europe and Western Asia, and another towards Denisovans in Asia. Neanderthals are characterized by a robust build, shorter stature than that of *Homo erectus* and *H. sapiens*, a protruding midface, with a wider pelvis and broader rib cage than modern humans.

As early *Homo* species entered the Middle Paleolithic, Neanderthals in particular exhibited skilled manufacturing of stone tools. This is evident by the Levallois method for creating sharp blade-like flakes, which is a specific technique of tool making within the Mousterian industry. This was a relatively complex stone tool industry where a stone core was prepared and flakes of a specific shape were removed for use as tools, including scrapers, projectile points, and burins. Neanderthals also made composite tools by using a combination of red ochre, a naturally occurring earth pigment, and bitumen, a component of asphalt, to make a resin-like adhesive. This was used to haft stone tools to wooden shafts for more complex toolmaking (Hoffecker 2018). Neanderthals were also known to use and control fire, as evident by hearths of fire-altered sediments containing charcoal, charred bone, and fire-cracked rock across several European and Levantine Middle Paleolithic sites (Aldeias et al. 2012; Henry 2017; Rosell and Blasco 2019; Leierer et al. 2020). Evidence suggests Neanderthals also used fire to create pitch from birch bark, another type of adhesive used to make composite stone tools. Neanderthal fossil sites are common in Western Europe and Western Asia, particularly in the well-watered river valleys of France. Fossil evidence has also been documented in the Middle East and Near East. Specific Neanderthal sites include Sima de los Huesos in northern Spain, Swanscombe in England, and Denisova Cave in Siberia. It is also generally thought that the southern peninsulas of Europe may have served as a refuge for Neanderthal populations during glacial periods.

The first fossils of Neanderthals were discovered in 1829 by Philippe-Charles Schmerling in the Engis cave, Belgium. Schmerling at the time did not recognize that the remains belonged to a distinct hominin species from modern *Homo sapiens*. These fossils consisted of a juvenile partial cranium (Engis 2), which was not identified as a Neanderthal until another discovery in 1856 at Kleine Feldhofer Grotte cave in the Neander Valley, Germany. A fossil assemblage of a skullcap, two femurs, various arm bones, ribs, parts of a pelvis, and a scapula belonging to Neanderthals was discovered at the cave by quarry workers. Geologist William King later proposed the name *Homo neanderthalensis*, for the fossil assemblages in 1863, after they were carefully examined by paleoanthropologists Johann Carl Fuhlrott and Hermann Schaaffhausen (Walker 2021). This was a crucial step in early paleoanthropology that not only confirmed the antiquity of the fossil assemblage but also established its distinctiveness from modern human remains.

Other fossil assemblages have been found at sites including the Shanidar Cave in Iraq and La Chapelle-aux-Saints in France, both of which have provided evidence that some Neanderthal skeletons at the site were intentionally buried (Dibble et al. 2015). This revelation has challenged the perception of inferior intellectualism and lesser cognitive ability regarding early hominins, suggesting that they were in fact capable of having deep and profound thought processes or insights that led to them carrying out coordinated actions. This is also evident in stone tool assemblages dating back to the Oldowan (Mode 1) and Acheulean (Mode 2) stone technologies found across Africa and Eurasia. Particularly noteworthy is the Shanidar Cave site, which has yielded 10 individuals below an Epipaleolithic-Paleolithic horizon along with evidence that some Neanderthals cared for one another (Spikins et al. 2018; Spikins et al. 2019). An adult male, Shanidar 1, was likely alive for 35-50 years despite evidence of a crushing fracture of his left eye socket, healed foot bone fracture in his right leg, and a withered right arm, further affecting his mobility throughout life. Remains of another adult male, Shanidar 3, had evidence of a healed rib injury from a projectile wound (Trinkaus and Zimmerman 1982). Shanidar 3's survival is likely attributed to being cared for by other members of its group, as he would have likely experienced significant pain and limited mobility. In total, four of the ten individuals, 1, 3, 4, and 5, represented by the fossil remains at Shanidar Cave, exhibited trauma.

Interestingly, the discovery of Shanidar 4 surrounded by clumps of pollen by archaeologist Ralph Solecki in 1960 led to the hypothesis that Neanderthals buried their dead with flowers, interpreted as evidence of ritualistic behavior (Solecki 1971; Leroi-Gourhan 1975). Recent excavations in 2015 and 2016 by archaeologists Tim Reynolds, Chris Hunt, and Graeme Barker into Solecki's main 14m-deep trench consisted of a four-by-four meter eastern extension, which investigated the Upper Paleolithic Baradostian cultural phase associated in the stratigraphy. There are four main chronological cultural phases identified in artifact assemblages from the site that have been radiocarbon dated, consisting of non-Levallois Mousterian stone tools associated with Neanderthals (>45,000 years ago), Epipaleolithic-Upper Paleolithic Baradostian (a regional variant of the Aurignacian stone tool industry) (33,000-27,000 years ago), Late Pleistocene Zarzian industry (12,000 years ago), a Proto-Neolithic cemetery (11,000 years ago), and Holocene activity (~7,000 years ago). Excavations of the Baradostian layers revealed primarily river pebbles, small cores and core shatter fragments, burins, and core edge-trimming flakes. Notably, the remains attributed to the individual Shanidar 5, estimated to be between 40 and 50 years old, consisted of fragmented vertebrae segments, a left wrist bone, a partial proximal left femur, a crushed partial pelvis, and the distal half of the right tibia and fibula, including their associated ankle bones. This discovery coincides with the initial remains of Shanidar 5 found in 1960, and the presence of burnt bone in faunal assemblages provided further consistency of a steppeland paleoenvironment with the initial palynology of the site, as they were from a species adapted to the cool and dry low-lying grasslands (Reynolds et al. 2015; Pomeroy et al. 2017).

Further investigations in 2016 revealed the remains of Shanidar Z, an elderly female individual that included a crushed skull, articulated upper torso assemblage, left shoulder blade, and right-hand bones. The discovery of Shanidar Z is significant pertaining to the pollen burial hypothesis as it was found within the previous assemblage of Shanidar 4, who is associated with initial palynology. It also supported the pattern of intermittently repeated burials that were carried out with coordinated efforts at Shanidar Cave by Neanderthals. Furthermore, the presence of a large, vertical rock within the burial site area may have been a landmark for these repeated burials (Pomeroy et al. 2020). While there is no doubt a substantial amount of evidence that Neanderthals buried their dead at the site of Shanidar, attesting to the capabilities for their empathy and intellectual coordination in similarity to Middle Paleolithic *Homo sapiens*, the evidence that pollen was associated with these contexts has been the subject of debate for the past several decades. Hunt et al. published a critical review of the pollen burial hypothesis to determine what the origins of the pollen within the burials of Shanidar 4 and Shanidar Z were, and if they were associated with such complexity regarding Neanderthal mourning practices. They found that while pollen clumps were found at these burials, further palynological analysis revealed that they contained pollen particulates from seasonally non-overlapping plants and were deposited by nesting ground bees within the site (Hunt et al. 2023). The study did not, however, contest the fact that there is evidence at Shanidar that Neanderthals did indeed bury their dead with intent, contributing to the ongoing debate surrounding Neanderthal cognitive abilities and cultural practices.

Significant and rapid shifts in the climate during the Last Glacial Period likely impacted the environments Neanderthals were acclimated to, changing the distribution of their food and resources within the ecosystem. While there is unequivocal evidence for Neanderthal DNA still present in modern human populations today, attesting to the idea that both *Homo sapiens* and Neanderthals interbred, early modern humans could have carried diseases that Neanderthals did not have the immunity to withstand. *Homo sapiens* are also often theorized to have outcompeted Neanderthals for resources. The exact reasons behind their widespread decline in population around 40,000 years ago, however, were likely a complex interplay between many of these factors that led to their physical disappearance, leaving residual Neanderthal DNA within some of us today.

"The Hobbit"

Homo floresiensis

100,000 to 60,000 years ago

While not considered a direct ancestor of *Homo sapiens*, *Homo floresiensis* likely evolved from a different early *Homo* lineage, potentially from a population of *Homo erectus* (or *Homo ergaster*) that became isolated on the Indonesian island of Flores between 1 million and 700,000 years ago (Aiello 2010; Dennell et al. 2014; Kaifu et al. 2024). Long-term isolation on Flores, where there were limited resources and reduced pressure from predation, likely led to island dwarfism in these populations. Island dwarfism is a type of phyletic dwarfism, also known as insular dwarfism, where the aforementioned conditions of an environment lead to reduced body size in large animals as a response to a lack of predators and isolated resources. The lack of predators particularly reduces selection pressure for large body size as a defense mechanism. This phenomenon is synonymous with the "Island Rule"; however, the degree to which this 'rule' can be applied universally is still debated in paleoanthropology and evolutionary biology.

H. floresiensis is characterized by its distinctly small stature compared to the rest of the genus *Homo* (roughly 3 feet 6 inches), a small skull with no prominent chin, and protruding brow ridges resembling traits of *Homo habilis* and Pliocene australopiths (P. Brown and Maeda 2009; P. Brown 2012). Their small stature earned them the nickname "The Hobbit," after the first fossils were discovered in 2003 by archaeologists Mike Morwood, Peter Brown, and Raden Soejono in the Liang Bua cave on Flores Island. The initial discovery was a nearly complete skeleton of an adult female (LB1), along with an isolated lower premolar (LB2) dated to approximately 100,000 and 60,000 years ago (Brown et al. 2004). Stone tools such as simple choppers and small flakes found alongside skeletal remains at the Liang Bua cave suggest *H. floresiensis* was capable of manufacturing stone tools comparable to the Oldowan tradition. These stratified artifact sequences span roughly 95,000 years with a minor shift around 11,000 years ago to unifacial flaking, a change in raw materials, and more frequent fire-cracked rock present (Moore et al. 2009). Evidence of earlier occupation on Flores by *H. floresiensis* was later discovered in stone tool and hominin fossil assemblages at the Middle Pleistocene site of Mata Menge, about 70km east of Liang Bua cave (Brumm et al. 2010). These were found in a fluvial sandstone layer that was deposited into a small valley stream around 700,000 years ago. This was concluded based on fission-track and argon-argon (40Ar/39Ar) dating on the volcanic ash strata which bracket the fossil deposit, combined with uranium-series coupled electron spin resonance

(US-ESR) dating (Brumm et al. 2016). This has provided substantial evidence that *H. floresiensis*, or a species of *Homo* ancestral to them, was present on the island of Flores long before it was initially thought. The paleoenvironment was likely characterized by the continued cooling and drying of the Middle Pleistocene, gradually shifting to mixed open grassland savanna-like habitats with a wetland component or riparian corridors.

"Ubag"

Homo luzonensis

148,000 to 50,000 years ago

On the island of Luzon in the Philippines, another species of archaic *Homo* was discovered in 2007 by an international team of researchers alongside archaeologist Armand Mijares and paleoanthropologist Florent Détroit. The initial findings of a single metatarsal (foot bone) fossil with unusual size and shape in Callao Cave raised many questions. This led to further excavations in 2011 and 2015, which uncovered more fossils in the same stratum of the cave. These included seven teeth, with five belonging to the same individual (holotype CHH6), a partial femur, along with hand and foot bones, initially dating between 67,000 and 50,000 years ago. This was later refined to an age spectrum of 134±14 ka for the fossils (Grün and Stringer 2023). The holotype fossil was attributed the name "Ubag," which was based on local mythology of a caveman in the region. These fossils exhibit a blend of australopith-like and modern *Homo* traits, which were found alongside cores, retouched chert flakes, and blade-like flakes dating back to the Late Pleistocene. The blend of both modern and primitive traits, with their curved fingers similar to some Pliocene australopithecines, suggests both bipedal and partially arboreal capabilities. *Homo luzonensis* also measures at around 4 feet tall, with some individuals potentially smaller than even *Homo floresiensis*. Some premolars of this hominin exhibit three roots, a feature uncommon in modern humans. This context led researchers to believe this was a previously unknown taxon, *H. luzonensis*, represented by these assemblages (Détroit et al. 2019).

An assemblage of 57 stone tools and butchered rhinoceros bones, alongside other fauna, was found in the Cagayan Valley of northern Luzon at the Kalinga site. While not found alongside any hominin remains, this suggests that an early hominin species, potentially ancestors of *H. luzonensis*, occupied the island much earlier than initially thought. These finds came from a clay-rich bone bed between 777,000 and 631,000 years ago (709±68 ka), dated using electron spin resonance (ESR) on the fluvial quartz deposits and tooth enamel of the fauna (Ingicco et al. 2018). Further geochronological studies of the Kalinga site using ESR on bleached quartz and argon-argon (40Ar/39Ar) single crystal laser fusion on plagioclase refined this range of human occupation presence to between 796±70 ka and 273±20 ka (Lambard et al. 2024). This provides substantial evidence of occupation periods and dispersal in Southeast Asia hundreds of thousands of years earlier than the *H. luzonensis* fossils were dated to, potentially multiple times throughout the Early and Middle Pleistocene. Recent research suggests this may have been similar to the insular isolation experienced by *Homo floresiensis*, with both groups potentially evolving from some *Homo erectus* groups that dispersed into various islands throughout Southeast Asia (Zanolli et al. 2022).

"Ju Lu"

Homo juluensis

300,000 to 50,000 years ago

The complex morphological diversity of Late Quaternary *Homo* species fossils discovered in eastern and Southeast Asia reflects the potential for variations to arise from multiple dispersal periods, rather than a single linear event into the region. A new taxon was recently proposed in 2024, *Homo juluensis*, that has raised questions about whether this will encompass certain Denisovan populations, specifically *Homo longi*, which may now represent Neanderthal populations (Fu et al. 2025a, 2025b). The Harbin skull, initially associated with *H. longi*, along with the Dali skull, which some have proposed as *Homo daliensis*, and the Jinniushan skull, have been called into question regarding their classifications as outliers to our understanding of Late Quaternary archaic humans. The representation of several distinct populations now by concrete fossil evidence in Southeast Asia, including *Homo floresiensis* and *Homo luzonensis*, has led to a more nuanced understanding of human evolution through dispersal periods and genetic interactions throughout the Late Quaternary. While this is still an active area of research, which includes paleoanthropology and genomic studies, it is still important to highlight the ongoing discoveries that contribute to this complex period of human evolution that led to the beginnings of Stone Age traditions in many parts of the world.

H. juluensis is derived from the Mandarin words "Ju Lu" meaning "large-headed," reflecting the remarkably large endocranial volume of this species between 1,700 and 1,800 cc. The taxon was proposed by paleoanthropologist Xiujie Wu and anthropologist Christopher Bae, with a central focus on the Xiahe mandible, Penghu 1 mandible (*Homo tsaichangensis*), and the fossilized remains from Xujiayao as the holotype and Xuchang as the paratype (Bae and Wu 2024). As discussed in the Denisovan section of this book, the Xiahe mandible from Baishiya Karst Cave was subsequently investigated by mtDNA analysis of sediment found within the site, revealing Denisovan presence around 100,000, 60,000, and up to 45,000 years ago (Zhang et al. 2020). ZooMS (collagen fingerprinting) coupled with proteomic analysis was also conducted on Middle to Late Pleistocene faunal assemblages from this site, which identified a hominin rib bone dating to approximately 48,000 and 32,000 years ago (Xia et al. 2024), consistent with the mtDNA analysis of Denisovan activity at the cave around this time.

The Penghu 1 mandible, initially proposed as *H. tsaichangensis*, was found off the coast of Taiwan in 2008 after a fishing vessel dredged it up from the Penghu Channel. It passed through the hands of collectors for several years until it was donated to Taiwan's National Museum of Natural Science in 2010. A 2025 study following the proposal of *H. juluensis* analyzed 4,241 amino acid residues from the mandible, which identified two Denisovan-specific variants. This concluded a wider distribution of Denisovans adapted to different climates than previously thought, supporting the findings of increased genetic and phylogeographic diversity among these populations (Tsutaya et al. 2025).

The Xujiayao fossil assemblage, initially attributed to archaic *Homo sapiens*, consists of twenty *Homo* fossils, 30,000 lithic artifacts, and roughly 5000 faunal remains attributed to mammals discovered in the Nihewan Basin of Northern China. It was recently dated to between 370,000 and 260,000 years old using electron spin resonance (ESR) of sediment quartz coupled with magnetostratigraphic analysis of the fluvio-lacustrine sequence occurring within the *Homo* fossil assemblage horizons (Ao et al. 2017). This not only revealed the horizon to be within the Brunhes Normal Polarity Chron, but represented the oldest Middle Pleistocene hominin fossils exhibiting Neanderthal derived traits, potentially of Denisovans, in East Asia.

At the Linjing Middle Paleolithic site in Xuchang County, Henan Province, central China, two fragmented partial skulls, Xuchang 1 and Xuchang 2, were excavated in 2007 and 2014, respectively. This assemblage also contained 15,000 artifacts and nearly 40,000 mammalian remains representing 20 taxa within an early Late Pleistocene stratum (layer 11) dated to approximately 125,000 and 105,000 years ago (Li et al. 2017; Zhao et al 2020; Wang et al. 2022). These faunal assemblages likely represent hunting activities, which were carried out by other Denisovans during roughly the same time evident by Middle-Upper Paleolithic tool assemblages in Denisova Cave (Shunkov et al. 2020a) and subsequent ZooMS analysis of bone fragments revealing percussive damage and meat exploitation (S. Brown et al. 2021). The Hexian fossil assemblage, including a skullcap and other skeletal fragments attributed to *Homo erectus*, was also proposed as part of *H. juluensis* reclassification by paleoanthropologist Sheela Athreya and anthropologist Yousuke Kaifu. This assemblage was discovered in the Longtan Cave in Hexian County, Anhui Province, China, and dates to approximately 412±12 ka years ago with uranium-series coupled electron spin resonance (US-ESR) dating, placing them well within the Middle Pleistocene (Grün et al. 1998).

The overwhelming evidence using mtDNA, proteomic, amino acid, and magnetostratigraphic analysis, along with ZooMS, U-series, and ESR dating, potentiates an impasse for the proposed classification of *H. juluensis*. It also raises questions about whether this reclassification will and should encompass Denisovan populations, considering our definition of them is ambiguous based on current fossil evidence. However, this highlights the greater degree of morphological characteristics and variation among the Late Quaternary hominin fossils of eastern and Southeast Asia, suggesting hominins of these regions may have also interbred rather than only undergone replacement by dispersals. This has become increasingly apparent given the taxonomic classifications among *Homo* species in this region, revealing greater diversity with distinct populations, such as *H. floresiensis* and *H. luzonensis*, contributing to a less unilinear picture of human evolution.

"Denisovans"

Homo denisova

285,000 to 25,000 years ago

Emerging during the Middle Pleistocene, Denisovans were a group of early humans that diverged from the lineage leading to modern humans about 700,000 to 400,000 years ago, and from the Neanderthal lineage about 473,000 to 381,000 years ago (Kuhlwilm et al. 2016; Zeberg, Jakobsson, and Pääbo 2024). However, as it stands today, they are considered to have been more closely related to Neanderthals, sharing a more recent common ancestor than either does with *Homo sapiens*. It should be noted that Denisovans are not officially classified as the taxon *Homo denisova*, or any taxa for that matter, as more conclusive evidence is required to establish species status. As such, Denisovans are the subject of a phylogeographic debate. Denisovans are known for several distinct characteristics, despite limited fossil evidence, including a wider pelvis, larger teeth, and wider, flatter faces compared to modern humans and Neanderthals. The evidence for Denisovans was first discovered at Denisova Cave in the Altai Mountains of southern Siberia, Russia. While unknown at first, DNA was extracted from a juvenile female finger bone in 2010 that had been recovered from Denisova

Cave, revealing a previously unrecorded archaic human lineage through genomic sequencing. This also allowed researchers to determine Denisovan distinctness from Neanderthals and modern humans.

A 2021 study revealed the mitochondrial DNA (mtDNA) analysis of 728 sediment samples in Pleistocene layers of the Denisova Cave, consisting of 685 ancient faunal mtDNA and 175 hominin mtDNA, with the earliest representing Denisovans. They are associated with the Early-Middle Paleolithic stone tool assemblages found at the site, deposited between 250,000 and 170,000 years ago, with the oldest securely dated remains being 200,000 years old (S. Brown et al. 2022). A turnover of Denisovan mtDNA at the site is observed that coincides with a faunal mtDNA composition change as well, suggesting a change in population around the gap period in their occupation of the cave between 130,000 and 100,000 years ago. It is likely that Denisovans and Neanderthals both occupied the cave repeatedly until the Upper Paleolithic, which is evident by the presence of modern *Homo sapiens* mtDNA appearing in the sediment layers around 45,000 years ago (Zavala et al. 2021). This represents one of the only known sites with evidence indicating occupation by Neanderthals, Denisovans, and modern humans (Z. Jacobs et al. 2025). Middle and Upper Paleolithic industry stone tool evidence discovered at Denisova Cave includes flakes detached from prepared cores, scrapers, and denticulate tools (Shunkov et al. 2020a). While their earlier manufacturing resembled Mousterian-like tools, by around 150,000 years ago, they began to use a more developed Levallois-like technique to produce narrow, parallel-sided blades. Later stratigraphy layers of Denisova Cave yielded an awl and eyed needle made from bone, and personal ornaments including a marble ring and a dark-green chloritolite bracelet dating to between 70,000 and 40,000 years old (Derevianko, Shunkov, and Volkov 2008; Shunkov et al. 2020b).

Another site known to be occupied by Denisovans is the Baishiya Karst Cave located on the Tibetan Plateau, where a mandible (Xiahe mandible) was discovered (Chen et al. 2019). Subsequent mtDNA analysis of sediment revealed Denisovan presence at the site 100,000, 60,000, and potentially 45,000 years ago (Zhang et al. 2020). A method from zooarchaeology known as ZooMS (collagen fingerprinting) was conducted on Middle to Late Pleistocene faunal assemblages from the Baishiya Karst Cave, identifying a hominin rib bone through further proteomic analysis dating to approximately 48,000 and 32,000 years ago (Xia et al. 2024). This was consistent with the prior mtDNA analysis, suggesting the bone was attributed to Denisovans occupying the cave, extending their presence here well into the Late Pleistocene. The same ZooMS analysis by Xia et al. also revealed that the modifications made to bone surfaces and other faunal assemblages indicated Denisovans were responsible for the exploitation of meat, marrow, hides, and bone tools. This was consistent with ZooMS-based analysis of 8253 bone fragments from Denisova Cave in 2021, which also revealed human-induced percussive means to fragment bones, including exploitation for meat (S. Brown et al. 2021).

Genomic sequencing analysis has revealed much about Denisovan genetic history and relationships with other hominins, including evidence that Denisovans interbred with both *H. sapiens* and Neanderthals (Ko 2016). This is reflected genetically in some modern populations, particularly in Papua New Guinea and other parts of Southeast Asia, exhibiting Denisovan ancestry ranging from 4-6% and others from 1-3% (G. Jacobs et al. 2019; Vespasiani et al. 2022). Recently, attention has been paid to an older discovery, the Harbin skull, otherwise known as the "Dragon Man," found by a laborer near the Songhua River in Harbin, China, around 1933. Its existence was unknown until 2018, when it was donated to Hebei GEO University following the laborer's death. Initially, it was attributed to the taxon *Homo longi* (Ni et al. 2021; Ji et al. 2021), but this classification was subsequently called into question due to its geographical location in northeast China, where other Denisovan evidence has been found and temporally overlapping with an estimated age of 146,000 years old. It also exhibited morphological characteristics resembling previously discovered assemblages, including a larger size, thick brow ridges, and large molars. This raised the possibility that the Harbin skull could, in fact, be the first nearly complete holotype skull specimen of

Denisovans. This hypothesis continued to gain traction until in 2025 mtDNA and proteomic analysis of the Harbin skull's dental calculus confirmed its Denisovan affinities, officially proposing it as belonging to the Denisovan population (Fu et al. 2025a, 2025b). This marked a significant turning point for understanding the morphological characteristics of Denisovans and their geographical spread amongst Neanderthals and *Homo sapiens*.

"Sapiens"

Homo sapiens

315,000 - 200,000 years ago to present

Humans did not evolve from monkeys. *We* evolved. While humans and monkeys are both primates, we more closely share a last common ancestor with apes than we do with any monkey. Humans, apes, and modern monkeys evolved along separate paths from that common ancestor, not by evolving directly into the other. Evolution is non-linear, branching with multiple different lineages adapting and changing independently of one another. We belong to our own distinct lineage of hominin mammals, whose divergence most likely began from the offspring of our last common ancestor with chimpanzees 6 to 8 million years ago.

The prevailing postulation for the origins of early *Homo sapiens* is the Out of Africa hypothesis, which puts forth that anatomically modern humans likely evolved in Africa from a population of *Homo erectus (Homo ergaster sensu lato*) or *Homo heidelbergensis sensu lato* (*Homo bodoensis*) between 315,000 and 200,000 years ago. One branch of archaic *H. sapiens* would have migrated out of Africa in multiple dispersal periods, with the first major migration around 70,000 to 60,000 years ago, potentially interbreeding or conflicting with Denisovans or Neanderthals, and populating the rest of the world. While the dispersal is often associated with this major event, more recent research is revealing that this may have occurred many times, each at its own caliber, possibly in smaller waves earlier than 100,000 years ago. Climate change, along with considerable environmental changes during the Last Glacial Period, drastically impacted faunal resources which are considered to be driving factors in the migration pattern out of Africa.

This has contrasted with the multiregional hypothesis, suggesting modern *Homo sapiens* evolved simultaneously from *Homo erectus* populations in different regions of the world, not just in Africa. It has been largely unsupported by genetic evidence focusing on mitochondrial DNA and Y chromosomes, revealing a more recent African origin for modern humans (Cann, Stoneking, and Wilson 1987). Mitochondrial DNA analysis used to trace human lineages has demonstrated that all currently living humans share a recent common matrilineal ancestor known as "Mitochondrial Eve." While she was not the first person alive, and certainly not the only female living at the time, the mtDNA of Mitochondrial Eve is the most recent female from which "all currently living humans descend through an unbroken line on their mother's side proves that modern humans only evolved once, most likely in Africa, sometime between 150,000 and 200,000 years ago" (Ko 2016). The multiple

Homo taxa represented in Southwest Asia have sparked controversy regarding whether some hominin populations ancestral to modern humans and Neanderthals might have emerged there during the Middle Pleistocene before the emergence of modern humans (Bermúdez de Castro and Martinó 2022). While not directly disregarding or rejecting the Out of Africa hypothesis on human origins, it highlights the possibility of a more complex evolutionary scenario having unfolded where the divergence of modern *H. sapiens* and Neanderthal lineages may have had a geographically diverse origin.

The earliest current evidence of anatomical *Homo sapiens* was found at the site of the Jebel Irhoud cave system in Morocco. This fossil evidence also supports the Out of Africa theory, and exhibits a mix of primitive and modern traits, indicating one of the first appearances of archaic *Homo sapiens*. However, it also suggests that we did not solely evolve in East Africa. The remains consisted of a skull, jawbone, teeth, and limb bones, which were dated using thermoluminescence to an age of 315±34 thousand years (Hublin et al. 2017). They were found alongside Levallois technique stone tools, indicating *Homo sapiens* emerged alongside sophisticated stone tool technologies in the Paleolithic (Old Stone Age). It was here that early modern humans began to use and control fire, improve stone tool technologies, and develop more elaborate clothing for warmth and protection. As the Stone Age continued throughout the Pleistocene, the adaptation of these technologies for survival may have been a response to the colder climate during the Last Glacial Period, which occurred between about 115,000 and 11,700 years ago.

The most widely accepted evidence for migration to the North and South American continents following the dispersal periods in Africa is through the Bering Land Bridge (Beringia), which existed approximately from 36,000 to 11,000 years ago after a drop in sea levels during the Last Glacial Period, between present-day Siberia and Alaska. Large amounts of water became locked into glaciers, facilitating movement over the revealed land bridge and perhaps long-term settlement on the landmass. It is often considered a major contributing factor to the first *Homo sapiens* migrating to the Americas. This migration marked the beginning of the Lithic/Paleoindian period throughout North, Meso, and South America until Early Archaic stone tool traditions emerged around 8000 BCE. However, the Bering Land Bridge has two main points of contention involving the route by which these populations migrated. The initial theory of a migrational route was formed on the basis of an "ice-free" corridor opening up between the Laurentide Ice Sheet and the Cordilleran Ice Sheet about 15,000 years ago through present-day western Canada and into the Pacific Northwest of the United States. This idea has been challenged by the Kelp Highway hypothesis, which suggests the migration primarily took place along the Bering Land Bridge and PNW coastline, where populations could rely on diverse resources within kelp forests. The Ice-Free Corridor (IFC) would have been a harsh, rocky environment with little foliage and fauna trapped between these two ice sheets, making survival over long migrational journeys immensely difficult. This coastal migration route was first proposed in 1979 by archaeologist Knut Fladmark, who argued that the greater abundance and stability of resources along the coast, such as the vast kelp forests, would have provided a more viable means of subsistence strategies (Fladmark 1979). It was later coined the Kelp Highway hypothesis by archaeologist Jon Erlandson in a 2007 study based on a model of the ancient Pacific coastline to determine when and where kelp forests once existed along viable migration routes (Erlandson et al. 2007). More recent cosmogenic nuclide dating of boulders left behind by the Laurentide and Cordilleran ice sheets also yielded an age of 13.8±0.5 ka for the Ice-Free Corridor, supporting that it was not fully open yet to facilitate the earliest evidence of migration into the Americas (Clark et al. 2022).

Formerly, the earliest evidence of Paleoindian occupation was an assemblage of artifacts discovered in 1929 from the Clovis type-site, Blackwater Draw, located in New Mexico. The full excavation of the site between 1932 and 1937 revealed fluted bifacial projectile points along with other Clovis-type stone tools, dating the Clovis culture to approximately 13,340 to 12,830 years ago (Boldurian 2008, 70). The discovery led to the initial Clovis First theory of the 1930s, which suggested this was one of the oldest known cultures in the Americas and directly ancestral to later Native American populations, providing the earliest benchmark stone tool technologies of occupation. This theory became intertwined with the Ice-Free Corridor hypothesis on the basis that Paleoindians would have migrated over the Bering Land Bridge while it was open for a limited amount of time at the end of the Last Glacial Period. Scholars were cautious to approach any pre-Clovis claims at the time, as these two theories presented it as impossible for there to have been an earlier occupation in the Americas. The Clovis First theory later became challenged and heavily scrutinized after excavations at Meadowcroft Rockshelter in Pennsylvania revealed prehistoric activity predating Clovis occupation. In 1973, archaeologist James Adovasio and his team uncovered thousands of distinctly pre-Clovis stone tools at the site, including flakes and projectile points, burned sediments, and pottery, along with human and faunal remains. The site was dated to between 19,000 and 16,000 years ago, significantly pushing the timeline back of human occupation in the Americas before Clovis (Adovasio 1978). From 1977 to 1985, further evidence of earlier occupation predating the Clovis culture was discovered at the Monte Verde site in Southern Chile by archaeologist Tom Dillehay. The site yielded remains of stone and wooden tools, faunal assemblages, and the remnants of a campsite with dates of human occupation around 14,500 to 12,500 years ago (Dillehay and

Collins 1988). Follow-up investigations at the Monte Verde site revealed a longer ephemeral human presence between 18,500 and 14,500 years ago based on stone tools, burned sediment, and faunal remains (Dillehay et al. 2015).

The original Clovis First theory of the 1930s came into controversy again with the discovery and dating of a Clovis burial. In 1968, construction workers at a ranch located near Wilsall, Montana, discovered the remains of an infant (Anzick 1) along with over a hundred lithic and antler artifacts covered with red ochre powder associated with the burial. The assemblage of stone tools was identified as Clovis-type and was likely buried intentionally with the infant, who was between one and two years old at the time of death. The site was uncovered while the workers were collecting sandstone gravel with a front end loader, and they proceeded to remove the bones and artifacts with excavators. While the burial was initially excavated without the supervision of archaeologists, further analysis of the assemblages revealed their association with the Clovis culture. However, there has been significant controversy surrounding the Anzick 1 site and its contents, given that it is the only known Clovis burial in the Americas, which has provided critical insight into their burial practices.

Initially, a 1991 study published the first radiocarbon dating for Anzick 1 using the hydroxyproline fraction of a bone specimen to 10,710±100 C14 BP (Stafford et al. 1991). The technique isolates this bone-specific amino acid from other molecules to remove contaminants, providing a more reliable and accurate age of a specimen. In 2007, a more refined age for the Clovis period as a whole was published in a study that revised the timeframe to between approximately 11,050 and 10,800 C14 BP. This timeframe reinforces evidence that populations were already living in the Americas predating the Clovis period by showing an overlap of Clovis with non-Clovis sites. (Waters and Stafford 2007). Based on this evidence, the Clovis culture had a much shorter and later existence than previously thought, suggesting it could have spread across North America in only 200 years while overlapping with non-Clovis sites. This makes it anthropologically implausible for Clovis to be the first culture to populate the entirety of both American continents in just a few generations. In 2014, Rasmussen et al. published the sequenced genome of Anzick 1, which found that the DNA is more closely related to Central and South Americans than to some North American groups (Rasmussen et al. 2014). This study reported a radiocarbon date of 10,705±35 C14 BP for the Anzick 1 individual based on XAD-purified collagen. The results support the theory of a much earlier migration into the Americas involving an ancient population that diverged into northern and southern branches, which predated the Clovis culture. These findings are consistent with the pre-Clovis presence in the Americas and also the dismissal of the Solutrean hypothesis, which suggests Native Americans originated from Europe.

However, there were still disagreements on the lithic tool assemblages coinciding with the radiocarbon dates for the Anzick 1 remains. To resolve debates about the lithic assemblages being considerably older than the burial remains they were found with due to contamination issues in earlier testing methods, a study in 2018 by Becerra-Valdivia et al. reassessed the chronology on the Anzick 1 site. The new testing method of isolated amino acids by using a more rigorous pretreatment in order to avoid contamination confirmed that the human remains and stone tools were part of the same synchronic burial (Becerra-Valdivia et al. 2018). This was calibrated to a date of 12,905-12,695 cal BP for the Anzick 1 site.

These studies found strong association with the Clovis culture based on the lithic assemblages, mortuary practices, and dating of the site, with the Anzick 1 genome confirming that Clovis was part of an ancestral lineage which populated the Americas from Asia, suggesting the lithic technologies spread throughout vast regions of the Americas, although it does not prove that the Clovis people were necessarily the absolute first inhabitants. The Anzick 1 genome has therefore refined our understanding to reflect that Clovis was a widespread and influential early culture, but that earlier populations also existed in the Americas. In 2014, Anzick 1's remains were returned in a reburial ceremony at Shields River Valley, Montana, that included members from several tribes in Montana, Washington, and Oregon. Many neo-diffusionist theories resulted from the archaeological and genomic findings of Anzick 1, which controversially argued that the genome instead supported the Clovis First theory of migration into North America. The genomic data from Anzick 1's remains were argued as directly ancestral to later Native American populations, supporting a single, rapid expansion from a proto-Clovis or Clovis population (Fiedel 2017, 2018). While these theories maintained that the genetic origin of Native Americans lay in Northern Asia, dismissing theories of European Solutrean origin, ongoing research suggests a more complex picture with multiple pulses of migration and admixture events occurring over a wider timeframe that predate the proposed timing for Clovis expansion. Archaeologists and geneticists criticized the Anzick 1 Clovis First interpretations as unsupported by broader scientific research, showing that the Anzick 1 genome actually supported a pre-Clovis migration and a more complex colonization history. The genetic data used actually helps support the now-dominant coastal migration theory, where Paleoindian populations moved south along the Pacific coast earlier than the opening of the Ice-Free Corridor, contradicting the strict interior-route Clovis First model entirely. In 2020, a follow up study by Michael Waters used radiocarbon dating, including dates from XAD-purified collagen from faunal bones of mammoth and bison, along with charred plant remains from ten different known Clovis sites, to narrow the overall timespan of the Clovis period (Waters, Stafford, and Carlson

2020). This process involved extracting collagen by dissolving the bone specimens and using a molecular sieve or ultrafiltration and decalcification to obtain purified amino acids for more accurate and reliable dating. The study concluded that the Clovis period was likely from around 13,050 to 12,750 cal BP, suggesting that the Clovis technology ended abruptly at the beginning of the Younger Dryas cooling event. This was a much earlier timeframe than the previous 2007 study placed the Clovis period, which also expanded the duration of their existence from 200 to nearly 300 years.

The date range of occupation in the Americas coinciding with the Ice-Free Corridor hypothesis was recently challenged once again by the discovery of well-preserved footprints within the Lake Otero outcrops, dating to between 23,000 and 21,000 years ago at White Sands National Park, also in New Mexico, which implicates the complete opening of the Ice-Free Corridor and earlier occupation (Bennett et al. 2021). These footprints were initially radiocarbon (C14) dated using *Ruppia cirrhosa* seeds found impressed into the sediment layers of the footprints. This was subsequently corroborated with further calibrated radiocarbon (C14) dating to confirm the age of the footprints using pollen found within the same stratigraphic horizons as the seeds. Researchers also used optically stimulated luminescence (OSL) dating of quartz grains in the surrounding sediment, which substantiated the evidence of earlier occupation through the coastline before the Ice-Free Corridor opened

(Pigati et al. 2023). This date range confirmed that humans had made it much farther south of the ice sheets in Canada before the end of the Last Glacial Maximum (LGM). It was also significant in proving that there was occupation in North America much earlier than the burial of Anzick 1.

From the earliest evidence of stone tool use beginning in Africa to the rise of Paleoindian cultures in the Americas, the prehistoric ages were already well on their way by the time *H. sapiens* emerged. The technological foundations set in motion by the hominins that came before us, and the ones living alongside us during this time, marked a significant turning point for the development of early cultures. Hunting megafauna and foraging in small bands of close cooperating members became part of early nomadic lifestyles, and stone tools became more specialized and refined for specific tasks. Early forms of art and music began to take shape, with spoken languages potentially existing over 100,000 years ago. Humans were forced to adapt to various climates and environments, including a glacial period, demonstrating our resilience and ability to innovate. This facilitated the development of fire for warmth, and soon after for control. Fire use became central to human health and survival throughout the Stone Age as it allowed us to begin cooking food and slowly establish more sedentary lifestyles towards the agricultural revolution. The Stone Age is the longest and most encompassing period of human prehistory, spanning millions of years, and some of the earliest traces of our cultural past begin here.

THE PREHISTORIC AGES

I have no mouth. And I must scream.

- Harlan Ellison

EARTH'S BEGINNINGS CAN BE TRACED BACK 4.5 billion years, yet human prehistory only accounts for a small fraction of its existence. The Prehistoric Ages, or when there was human and early-human ancestor activity before recorded and documented human history, roughly dates back to 3.3 million years ago and coincides with the first evidence of stone tool use. It is generally categorized into three archaeological periods: the Stone Age, Bronze Age, and Iron Age, and sometimes into different eras.

Our prehistory began with the appearance of the oldest stone tools found at the Lomekwi 3 site in Kenya, dating to around 3.3 million years ago (Harmand et al. 2015). The discovery suggests stone tool use predated the *Homo* genus by nearly 700,000 years, potentially including species such as *Australopithecus afarensis* or *Kenyanthropus platyops*. Their association with early stone tool use marks the beginning of the Stone Age

Paleolithic Era. From the invention of stone tools made for simple functions, advances in food production and agriculture, to early examples of art and religion, prehistory's enormous time span, ending roughly 3,200 years ago (dates vary upon region), was a period of great transformation on Earth.

The Paleolithic is considered an era even though it occurred during the Pleistocene Epoch because the term *era* in this context refers to a distinct period in human history. This is usually characterized by specific technological advancements, namely the widespread use of rudimentary stone tools, which aligns closely with the time frame of the Last Glacial Period. Essentially, the Paleolithic Era is the "Old Stone Age" within the larger geological time scale from the Pliocene through Pleistocene epochs.

While *epoch* is a geological term signifying a specific period within a larger geological era, the term *era* in archaeology often refers to a distinct cultural period marked by significant technological developments. In human history, eras are typically defined based on major cultural or societal shifts, like the Medieval Era or the Industrial Era. The Geologic Time Scale and International Chronostratigraphic Chart refer to eras being defined by significant changes in Earth's life forms and geological features, often marked by mass extinction events. They are typically measured in millions of years, geologically speaking, making them vastly longer time periods than human historical and prehistoric eras.

THE STONE AGE

3.3 million years ago to 2000 BCE

The Stone Age is perhaps one of the most evident archaeological ages globally, as the use of copper, bronze, and iron technologies in the Americas was not nearly as widespread as in Eurasia until much later. It began with the first evidence of stone tool use associated with our early hominin ancestors around 3.3 million years ago in Africa. These were simple choppers and crudely worked handaxes used as subsistence strategies; however, their usage by early hominins underscores a critical period in prehistory when our ancestors began the long constitutive relationship with stone tool technologies.

For a time during the Stone Age, early humans shared the planet with several extinct hominin relatives, including Neanderthals and Denisovans. This includes earlier *Homo* species, like that of *Homo erectus*, which existed on Earth for nearly 2 million years, a disproportionately long amount of time compared to our own species' existence. The coexistence of multiple hominin species on Earth throughout the Lower Paleolithic brought gradual innovation and shared use of different stone tool technologies together as populations not only encountered one another through migrational routes, but also adapted their tool typologies through the determinism of stone material and its reactivity to percussion methods of toolmaking. Thus, the properties of desirable stone material, along with its environmental accessibility, impart a symbiotic relationship with hominin and early human subsistence strategies.

By the end of the Stone Age, *Homo sapiens* were the dominant and only hominin species left on Earth. The development of sedentary and agricultural lifestyles from cultures of hunting and gathering soon became more widespread in the Neolithic Era, influencing the development of cultural traditions and interconnected populations of humans. The Stone Age is divided into three separate eras, each with its own subperiods: the Paleolithic (Old Stone Age), Mesolithic (Middle Stone Age), and Neolithic (New Stone Age). The following sections will briefly cover all three eras and their subperiods as a general reference. It is important to keep in mind that the Stone Age began at vastly different times around the world, as early humans did not exist in the Americas until migration during the Late Pleistocene. Africa was a center for early Stone Age development as the birthplace of stone tool use. The earliest forms of percussive techniques were developed by hominins as a means to process food. By around 1.8 million years ago, *Homo erectus* migrated out of Africa, which coincided with the dispersal of Oldowan and the more sophisticated Acheulean stone tool technologies throughout Eurasia in the archaeological record.

Lower Paleolithic

3.3 million to 300,000 years ago

The Lower Paleolithic marks the very beginnings of stone tool use among early hominins, and it is the longest subdivision of the Old Stone Age. Stone tool technologies did not all emerge at one point in time, and they instead were gradually adopted as early hominins spread out from Africa and eventually became replaced by *Homo erectus* and *Homo sapiens* until the Upper Paleolithic, when our own species remained the last surviving of the *Homo* genus, usurping Neanderthals and Denisovans by as late as 40,000 to 15,000 years ago. Different regions around the world entered their own stone ages at vastly different times, and these entries are intended to juxtapose the earliest use of stone tools in the archaeological record with their widespread adoption globally throughout the next several million years. Lower Paleolithic stone tool technologies are characterized by large and small scrapers, hammerstones, choppers, awls, and other early simple stone tools. By the later part of the Lower Paleolithic and into the Middle Paleolithic, early humans began to control fire for warmth, protection, and cooking while living in small nomadic groups. While frequently relocating, they often resided temporarily near permanent water sources, as stone tool material was abundant in river valleys.

The Lower Paleolithic is often studied in two stages based on stone tool industries: the Oldowan Stage (2.9 million to 1.7 million years ago) and the Acheulean Stage (1.7-1.5 million to 250,000-200,000 years ago). The Oldowan industry, discovered in Olduvai Gorge, Tanzania, is characterized by simple chipped or worked rock used for cutting, scraping, and chopping. Acheulean stone tools, first discovered in Saint-Acheul, France, typically exhibit more refinement with bifacial handaxes and cleavers, as they appeared much later than Oldowan technologies. While the Oldowan tradition formerly was considered one of the earliest known stone tool technologies, often attributed to *Homo habilis*, *Paranthropus boisei*, or even *Australopithecus garhi*, the discovery of stone tools at the Lomekwi 3 site near Lake Turkana, Kenya, dates to around 3.3 million years ago, which may warrant classification as the Lomekwian Stage (Harmand et al. 2015). These stone tools are thought to have potentially been developed by earlier species of hominins, with candidates including *Australopithecus afarensis* or *Kenyanthropus platyops*. They are characterized as simple, larger-sized cores with flakes struck off from an anvil surface. This discovery revealed that other early hominin species were perhaps manufacturing and using stone tools 700,000 years before the genus *Homo*.

Middle Paleolithic

300,000 to 50,000 years ago

The Middle Paleolithic brought anatomically modern humans into the foreground of the Pleistocene as they began migrating out of Africa at the end of the period and into the Upper Paleolithic, likely in multiple waves of dispersal throughout Europe and Southwest Asia. These populations would have interbred and also replaced groups of Neanderthals and Denisovans in Eurasia, leading to the widespread population of the world by *Homo sapiens* throughout and after the Upper Paleolithic. The Middle Paleolithic, which largely began in Africa, Europe, and Asia, is characterized by significant adaptations in stone tool technology from the preceding period. This included the appearance of the Mousterian and Tayacian stone tool industries in the archaeological record, representing a major step in early human cognitive evolution. Mousterian industry stone tools are characterized by projectile points, scrapers, and hand axes. The type site of the Mousterian industry is located at Le Moustier cave in the Dordogne region of France. Stone tools discovered at the site consist of Levalloisian cores, Acheulean bifacial cores, and a discoidal core flint tool industry associated with the Clactonian tradition, indicating a rich site use history.

The Levalloisian technique, a method of the Mousterian industry, involved carefully preparing a stone core to produce flakes of a predictable size and shape for various functions. This innovative technique led to more effective tools for hunting and processing food, which in turn aided in the mobility of these populations in ephemeral environments where fauna were seasonally transient. It also required more abstract thought, planning, and foresight in order to work a stone core that would produce these longer, more predictable flakes. While other hominin species have been known to use this technique, it is often associated with Neanderthal stone tool manufacturing. Neanderthals are also known to have "mixed pine resin with beeswax and bitumen with quartz and gypsum, and distilled tar from birch bark" (Doronicheva et al. 2022, 17421) to form adhesives for making composite stone tools. This higher level of technological complexity with Neanderthal populations challenges older and outdated views of Neanderthals as less cognitively capable than *Homo sapiens*. The creation of composite stone tools presents evidence that Neanderthals were, in fact, capable of premeditated thought processes, a sophisticated understanding of stone materials, and cultural development.

Around 110,000 years ago, the final populations of *Homo erectus* began to vanish following their gradual decline throughout the Lower and Middle Paleolithic. By the end of the Middle Paleolithic, small nomadic groups of early humans had already begun to develop simple social structures. The use of fire had also become widespread by this time among both early humans and Neanderthals following its control and exploitation by *Homo erectus*. While other early hominins may have taken advantage of natural fires, it was *Homo erectus* that was able to maintain and master the controlled use of fire. This facilitated greater social adaptations and cultural development by providing sources of warmth and light, along with the ability to heat and prepare food, promoting a dietary shift in the genus *Homo*. Additionally, the controlled use of fire changed composite tool technologies as it provided enhanced toolmaking capabilities, such as the production of adhesives and fire tempering stone tools. This process made workable stone material more rigid and brittle by altering its molecular structure through rapid cooling, which creates microscopic fractures under the internal stresses within the rock. Stone heated in this way is easier to chip and flake, allowing for the fabrication of much sharper and more refined tool technologies.

Symbolic practices among surviving Neanderthal populations and *Homo sapiens* became prominent in the Middle Paleolithic, including intentional burials of the dead, the use of pigment, and the creation of personal ornaments reflecting the development of shared belief structures, complex social networks, and cultural identity among different groups. While these social systems preceded the complex agricultural societies of the Neolithic, they were crucial adaptations that allowed groups of early humans to survive across diverse environments as they migrated out of Africa and into Eurasia. Neanderthal and Denisovan populations met a decline by the end of the Middle Paleolithic, with the last groups on Earth mostly disappearing by the beginning of the Upper Paleolithic.

Upper Paleolithic

50,000 to 10,000 years ago

The Upper Paleolithic saw the most significant cultural developments of the Paleolithic Era, despite it only being a fraction of this subperiod as a whole. Early drawing and art-form techniques, including cave paintings, emerged just before the beginning of this subperiod. The oldest known figurative rock art was found at the Leang Karampuang cave on the island of Sulawesi, Indonesia, in 2017. According to the publication on the discovery and its dating analysis, it was "Painted at least 51,200 years ago, this narrative composition, which depicts human-like figures interacting with a pig, is now the earliest known surviving example of representational art, and visual storytelling, in the world" (Oktaviana et al. 2024). As small societies began to develop these intricate social patterns, the use of masks and simple musical instruments also indicates the movement towards more complex ceremonial practices among Paleolithic cultures.

Stone tool industries became more complicated with many earlier hand-axe and simple flake tool technologies being replaced by specialized blades and retouched bladelets. These tools were struck from specially prepared cores that allowed for long parallel-sided flakes, known as a pre-form, to emerge from the bulb of percussion at the striking angle. Projectile point hafting became more complex from earlier Neanderthal adhesive techniques, and more sophisticated composite tools like harpoons and atlatls are estimated to have also become more widespread. Bone, antler, and flint were also incorporated into flint knapping techniques and composite toolmaking. Animal hide clothing, along with tools like burins and needles in combination with thread, was more common during this time.

By the end of the Upper Paleolithic, the Bering Land Bridge that had enabled the migration of humans to the Americas had begun to disappear. The Last Glacial Period of the current ice age had reached its near end, soon facilitating agricultural and sedentary lifestyles among populations as global climates warmed towards the Neolithic Era. As conditions became more favorable for farming and the domestication of animals, settled communities began to emerge. Different species of early hominins also started to decline in populations, including Neanderthals, as early modern humans rose to predominance.

Epipaleolithic

25,000 to 10,000 years ago

The Epipaleolithic is considered to be the final stage between the Paleolithic and Mesolithic Eras, marking the end of the Paleolithic as a whole in some parts of the world, with a focus primarily in North Africa, the Levant and Near East, the Caucasus region, and Southwest Asia. This period is typically defined as occurring around the end of the Last Glacial Maximum (LGM), when expansive ice sheets began to melt, facilitating the movement of human populations and new fauna into formerly inaccessible regions. However, it also restricted transcontinental migration by the rapid marine transgression of exposed land bridges throughout the Late Pleistocene and Early Holocene following the LGM. Hunter-gatherer populations were met with the influence of post-glacial warming and inundation of coastal settlements, which presented new possibilities for cultural and technological adaptations within inner coastal woodland areas. Stone tool technologies of the Epipaleolithic are characterized by the use of small stone tools called microliths and nomadic lifestyles among groups of humans. Humans began making more refined tools from small flint and obsidian blades that were hafted into wooden implements. Pottery also began to emerge in different regions of East Asia and eastern Russia during the Epipaleolithic.

The timing of the Epipaleolithic corresponds to the Lithic/Paleoindian period in the Americas, with the first evidence of Paleoindian occupation dating back to around 25,000 to 12,000 years ago in different geographic regions. The Bering Land Bridge, which existed between present-day Siberia and Alaska, was a contributing factor in the first humans migrating to the Americas and began to disappear towards the end of the Paleolithic Era. As massive ice sheets and glaciers began to melt following the end of the Last Glacial Period around 11,700 years ago, sea levels rose, forests spread replacing many open steppelands, and populations of megafauna started to dwindle. Due to these factors, adaptations to foraging, hunting smaller game, and a reliance on early fishing methods became incorporated into daily life.

Paleoindians of the Americas are known to have temporarily resided in well-drained environments near permanent water sources like rivers and lakes, where stone tool material was abundant. Early Paleoindian stone tools are characterized by bifacial projectile points, scrapers, and flakes made from chert, flint, chalcedony, quartz, and jasper. Many Paleoindian cultures became regionally distinct, with Folsom in the plains, Gainey in the Great Lakes, Midland in the Midwest, and Suwannee in the Southeast of North America. The Clovis culture (ca. 11,100 to 10,800 BCE) is one of the more well-known Paleoindian stone tool traditions, exhibiting finely worked fluted projectile points and retouched blades. They resided across North America, extending from southern Canada to northern Mexico. While Clovis is considered an early Paleoindian culture, pre-Clovis stone tools discovered at the Meadowcroft Rockshelter site in Pennsylvania, including flakes and projectile points, burned sediments, and pottery, along with human and faunal remains, suggest an earlier human occupation between 19,000 and 16,000 years ago (Adovasio 1978). Other notable sites that have yielded evidence predating the Clovis culture include the Monte Verde site in Southern Chile and the Lake Otero Outcrops in White Sands National Park, New Mexico. Human occupation at Monte Verde dates to between 18,500 and 14,500 years ago (Dillehay and Collins 1988; Dillehay et al. 2015), with footprints at White Sands dating to between 23,000 and 21,000 years ago (Bennett et al. 2021). These discoveries significantly pushed the timeline of human occupation back in the Americas, and further excavations continue to contribute to our understanding of the Lithic/Paleoindian period.

Early Mesolithic

10,000 to 7700 BCE

The Early Mesolithic is also known as Mesolithic-I and corresponds to the Preboreal period in Europe, a time of significant warming trends following a pulse of abrupt cooling at the end of the Last Glacial Period known as the Younger Dryas. At the end of the Paleolithic Era, a landmass connecting Britain and continental Europe, known as Doggerland, was exposed as water became locked in ice sheets during the Last Glacial Period and the Younger Dryas. This landmass facilitated the movement of Mesolithic cultures and was also inhabited by permanent and ephemeral settlements.

A 2024 study found that the inundation of Doggerland may have begun rapidly between 10,500 and 10,000 years ago based on comparative radiocarbon date models (Hoebe et al. 2024). While sea level rise caused significant land loss, some areas, such as northwest Germany and the Netherlands, experienced increased hunter-gatherer activity. This suggests that the impact of marine transgression on coastal settlements did not simply cause a unilinear decline and population displacement. Rather, this may have prompted cultural change and shifts in activity, with the relationship between inundation of Doggerland and human occupation not being strictly uniform across the landscape. However, the distribution of archaeological data collected is limited by preservation and accessibility, especially in submerged contexts. This problem continues to complicate associations between Early Holocene eustatic sea level rise and human behavior across landscapes. The Doggerland landmass gradually became further inundated during the Early Holocene and was nearly fully submerged by approximately 8,200 years ago at the beginning of the Mesolithic.

The Mesolithic subperiods were a time of transition from the Paleolithic to the Neolithic Era, primarily associated with Europe and Western Asia. The Early Mesolithic is characterized by the gradual shift from a hunter-gatherer to a semi-nomadic way of living, the development of microliths, and the beginning of animal domestication in many geographical regions of Europe following the Upper Paleolithic. As Europe began to experience warmer climates in the Preboreal period following the end of the Last Glacial Period, the spread of forested ecosystems would dramatically shape human development in the Mesolithic. The rapid increase in average annual temperatures dramatically reshaped ecosystems in the southern Baltic region (Schmölcke et al. 2006), with Scots pine and birch becoming dominant among spreading forests. These species of trees were accompanied by hazel, elm, and oak soon after, providing forested habitats for red and roe deer, wild boar, moose, aurochs, brown bears, European otter, beavers, and foxes (Jöns and Harff 2014, 178). Temperate wetlands also replaced many of the mammoth steppes in Eurasia that the remaining megafauna of the Pleistocene relied on as seasonal habitats.

Early Mesolithic stone tools were small, sharp, and carefully crafted microliths. They were often made from flint and were used for hunting, fishing, and food gathering. These microliths were typically triangular or trapezoidal, being meticulously carved into their desired shape and size. Chip-stone axes were made from large pieces of flint that were broken into shape. Early axes were rough, but later Mesolithic axes had more refined surface flaking. Retouched bladelets also became prominent, which were small chipped stone tools that were used to create composite hafted tools. Other stone tools of the Early Mesolithic include oblique blunted blades, also called pen-knife points, which were made from a small flake or narrow blade, as well as tranchets, which were a small flake tool with a single straight cutting-edge.

Middle Mesolithic

7700 to 5500 BCE

The Middle Mesolithic is also known as Mesolithic-II and corresponds to the end of the Preboreal period in Europe. Warming trends that followed the end of the Last Glacial Period brought a significant eustatic rise in sea levels, which reshaped many coastlines of Europe, Great Britain, and the Baltic region. While this pushed many groups inland, it also opened new corridors that were once covered by ice sheets for populations to follow migrating herds. The foundations of Neolithic agricultural practices became more evident with shifts in food sources, with permanent settlements that relied on fishing, plant cultivation, and early forms of animal domestication.

As humans started to trend towards more settled lifestyles with less nomadic movement, there was an increased reliance on plant foods. Although not as well established as the following Neolithic periods, the beginnings of small-scale agriculture allowed people to exploit local resources more efficiently, leading to a greater understanding of plant life cycles in temperate climates. While still primarily relying on hunting and gathering, human populations were able to compensate for larger settlements with the means of simple agriculture. Humans still depended on inland hunting and foraging, where seasonal environmental changes provided different distributions of abundant natural resources and fauna. This supplemented the trend towards early agriculture practices of the Neolithic Era.

Microlith toolmaking techniques became more refined with a focus on creating specialized tools for specific tasks like harvesting plants and fishing. Composite hafted tools using bifacial projectile points, crescents, and backed blades were used to process and refine animal hides, with awls being used to pierce leather and wood. Smaller bird or jewel points of the Mesolithic, often tanged or serrated with sophisticated bifacial flaking, indicate a shift in reliance on the bow and arrow for hunting game. A 2025 study found that microlith shape in plan-view did not impact their performance as projectile points. Instead, their form, thickness, and hafting technique were the variables that directly correlated to this factor, and their geometric variations are more likely culturally related (Cortell-Nicolau, Key, and Palomo 2025). This supports the idea that microliths can be used as indicators of Mesolithic presence patterns throughout Europe.

Late Mesolithic

5500 to 4000 BCE

The Late Mesolithic subperiod marked a transition into Neolithic tool and agricultural practices, while some communities still retained their Mesolithic heritages. The gradual and uneven adoption of animal husbandry and agriculture in different regions of the world largely contributes to the overlap in date ranges for Mesolithic and Neolithic period traditions. The construction of large stone structures known as megaliths associated with burial sites became prominent during the Late Mesolithic. The widespread construction of megaliths marks an important transition to the agricultural characteristics of Neolithic cultures. These Late Mesolithic and earlier Neolithic tombs were made of large stone blocks, the oldest of which were constructed along the Brittany coast of Northwest France, dating between 4700 and 4500 BCE, and in Antequera, Spain, dating to between 3800 and 3600 BCE. Although Neolithic megalith structures predating these tombs have been discovered at Göbekli Tepe in Turkey, dating to roughly 9500 BCE, the megaliths in France and Spain are considered the first megalithic grave structures, known as dolmens, which were covered stone tombs.

Towards the end of the Mesolithic Era, some populations began growing annual crops and practicing animal husbandry, which further distinguished societies of the Early Neolithic. Composite stone tools generally became smaller, more standardized, and increasingly focused on refined microlith technology. In Britain, new microlith technologies appeared, including scalene triangles and narrow, straight-backed rods, which were retouched along one edge (Momber 2014, 197). These were still hafted onto wooden tools like arrows and spears, demonstrating a continued shift from the Middle Mesolithic towards more complex composite microlith tools for hunting and processing food. This likely reflects adaptations to the changing global environments of the period, and the general warmer climate after the end of the Last Glacial Maximum (LGM). As the last isolated populations of mammoths began to die out around 4,000 years ago, eventually leading to their extinction, the adaptation to a warmer post-LGM climate pushed human populations to hunt and process smaller game with an increased reliance on supplemental fishing and annual agriculture. While microliths were present in the Middle Mesolithic, the Late Mesolithic saw a significant increase in their production and use with more refined shapes of backed blades, crescent lunates, and trapezes becoming incorporated into stone tool technologies of Mesolithic cultures.

Pre-Pottery Neolithic A

10,000 to 8800 BCE

The Pre-Pottery Neolithic refers to the early stages of the Neolithic Era, where agriculture and settled communities were present; however, pottery technology was not typical or widespread, even if pottery existed in other regions at an earlier time. It signifies a specific phase within the Neolithic Era where pottery was not a common practice in the Near East, even if a few isolated instances of earlier pottery might exist elsewhere. The Pre-Pottery Neolithic A (PPNA) has a geographical focus in the Levant and Upper Mesopotamia. Jericho, considered one of the oldest continuous settlements in the world, appeared during this time, around 9,500 to 9,000 BCE. The Pre-Pottery Neolithic A succeeds the Natufian period of the Mesolithic and Epipaleolithic Near East, characterized by the development of nomadic hunter-gatherer cultures to semi-sedentary lifestyles of the Pre-Pottery Neolithic, where animal domestication and early agriculture began to emerge in the region. Stone tool technologies during this time typically included flaked lithics like sickle blades, projectile points, transverse-blow axes, polished adzes, and specialized blades struck from regular cores, reflecting the transition to an agricultural and sedentary lifestyle in the Near East. These tools were designed for harvesting crops and processing grains, alongside continued use of projectile points similar to the late Natufian culture of the Near East.

Some of the earliest known pottery predating the Pre-Pottery Neolithic was discovered at the Xianrendong cave system in the Jiangxi Province of China, dating back to 20,000 years ago (Wu et al. 2012). It provides evidence of early human hunter-gatherers using pottery for cooking long before many cultures of Europe entered their agricultural shift into the Neolithic. This discovery predates the megalithic structure site of Gölbekli Tepe in Turkey, where over 600 fragmented vessels made from stone used at the Pre-Pottery Neolithic settlement have been uncovered in ongoing excavations since the mid-1990s (Dietrich et al. 2020). The creation of fired pottery transformed early human settlements by enabling easier transport of food, water, and natural resources. It also established new methods of cooking and processing plant goods. Pottery allowed agricultural products to be sealed away during leaner seasons, where grain, seeds, and other materials could be stored securely preventing them from becoming unstable. The widespread adoption of pottery-making throughout many regions of the world later in the Neolithic Era would ultimately shape how early human agricultural societies functioned, as many parts of the Near East and Europe entered the Copper and Bronze Ages.

Pre-Pottery Neolithic B

8800 to 6500 BCE

The Pre-Pottery Neolithic periods both represent the early stages of Neolithic cultures in the Near East, where people were just beginning to transition from hunter-gatherers to a more settled agricultural lifestyle. Pottery-making was still considered less widespread in many regions during the Pre-Pottery Neolithic B (PPNB), which is a defining characteristic of the later Neolithic period. While there is some overlap in dates for what is generally considered the Mesolithic and the start of the Pre-Pottery Neolithic, this is geographically based on different regions and cultures incorporating Neolithic agricultural practices, animal domestication, and animal husbandry. The geographic distribution of PPNB sites include parts of Western Asia and more regions of the Near East centered around the Levant and Upper Mesopotamia, which eventually impacted the development of agricultural practices in Europe.

While hunting remained an important part of subsistence, the major differences observed in PPNB from PPNA sites are the increased reliance on domesticated animals, along with changing architectural styles and social structures centered around agriculture, evident in the need for more substantial communities. Stone tool technologies from this subperiod included large tanged points, sickle blades, scrapers, and burins. The bidirectional blade technology used to make larger tanged points originated in the Middle Euphrates Valley at the beginning of the PPNB around 8900 to 8800 BCE (Borrell 2017). These tools were often made from flint, and many of them were hafted either transversely or laterally, depending on the intended purpose.

The Neolithic Revolution

10,000 to 2000 BCE

The Neolithic Revolution, or agricultural revolution, does not necessarily represent a specific period of time, but rather the emergence of Neolithic practices in Mesolithic cultures of the Near East, which gradually impacted and also developed independently in other cultures throughout Western Asia and Europe over several thousand years. The term "Neolithic Revolution" was coined by archaeologist Vere Gordan Childe in his 1936 book *Man Makes Himself*, where he draws on comparisons to the blossoming of agricultural practices during this time to the Industrial Revolution's technological milestones as a watershed moment in human history (Childe 1936, 66-104). Much of the Neolithic Revolution encompasses the Pre-Pottery Neolithic subperiods in the Near East and Europe, with the last 4,000 years of the Neolithic spanning this major shift in human settlement lifestyles in other regions of the world. Throughout the Early, Middle, and Late Mesolithic subperiods, the gradual shift to agricultural practices, animal domestication, and sedentary lifestyles among communities ignited the Neolithic Revolution in the Fertile Crescent.

The Pre-Pottery Neolithic A and B periods overlap with the Mesolithic, as this time was considered a transitional phase between the hunter-gatherer lifestyles of the Paleolithic and the early farming practices of the Neolithic. The Pre-Pottery Neolithic represents the initial stages of agriculture development in the Near East directly following from the Mesolithic Natufian culture. The latter half of the Neolithic Revolution spans the Pottery/Late Neolithic Period. The Neolithic Revolution is characterized by the beginning of crop cultivation, domestication of animals, and a shift to permanent villages, along with the widespread incorporation of pottery. The Neolithic cultures of the Near East and parts of Europe gradually entered their own Copper Age with the early advent of malachite reduction to extract raw copper. In Central Europe, the Linear Pottery Culture represents one of the earliest farming societies in the region, present from roughly 5500 to 4500 BCE. These communities utilized enormous longhouses and are recognized for their distinct pottery techniques decorated with linear motifs.

In North America, Paleoindian cultures such as the Plainview and Firstview represented post-Folsom projectile point technologies in the Great Plains. While not a uniform transition in the Americas, Paleoindian cultures entered the Archaic Period around 8000 BCE, which lasted until about 1000 BCE. It is commonly divided into the Early, Middle, and Late Archaic subperiods. Early Paleoindian cultures shifted from ice age hunter-gatherers relying on some of the last megafauna like mammoths and mastodons, to a more sedentary lifestyle during the Early Archaic as the climate continued to warm, sea levels rose, and forests spread further over new areas of land following the Last Glacial Period. Settlements of longer-term occupation in sites near permanent fresh water sources became increasingly common for bands of twenty to fifty people. The presence of chert, jasper, flint, and quartz, along with their varieties near riverbeds, provided Archaic cultures with abundant stone tool material. Archaic stone tools include composite bifacial projectile points and knives (PPK), grinding stones, net sinkers, scrapers, simple flake tools, and hammerstones. In the Great Lakes region of North America, evidence of Archaic period cultures mining and working copper is associated with the Old Copper Culture emerging around 9,500 years ago and peaking between 7,000 and 5,000 years ago before the use of copper began to decline (Pompeani et al. 2021). These early users of copper ore fashioned tools, weapons, and ornaments, demonstrating technological prowess and innovation among many Archaic cultures of the time. The use of copper in the Americas predates many regions in the Near East and Europe that entered their Chalcolithic Era, or Copper Age, thousands of years later.

While pottery-making in Archaic cultures was not yet a widespread practice, the earliest isolated instances in South America date between 7,500 and 5,000 years ago and were discovered in the Amazon Basin at the site of Pedra Pintada Cave, Brazil (Roosevelt et al. 1996). The presence of earlier pottery-making traditions in the Amazon Basin suggests some Paleoindian and Early Archaic cultures adapted to better resource management, without having to rely solely on big-game hunting. This may have been through fishing or foraging for nuts and fruit. In North America, some of the earliest evidence of pottery dates to roughly 4,500 years ago, discovered on Stallings Island, Georgia (Sassaman 1993). Pottery of the Late Archaic Stallings culture was often fiber tempered, with clay containing plant material like Spanish moss or palm leaves to reinforce the durability of vessels after firing.

By the end of the Archaic Period, the use of copper tools in North America had slowly declined while pottery-making began to transform horticulture practices in early Woodland Period settlements. Copper became a common material for ornamental purposes due to its malleability and relative softness, with copper tools being replaced by stone as it was more readily available and valued for its durability. Monumental Late Archaic earthwork settlements like Poverty Point, located in northeastern Louisiana, began development around 1700 BCE. It was later completed around 1100 BCE. Constructed by the Late Archaic Poverty Point culture, the impressive large earthenwork settlement suggests it was used as a central hub for trading, ceremonial purposes, and possibly for permanent residency, demonstrating the elaborate social organization surrounding the site. At its peak, the settlement is estimated to have had a population of 4,000 to 5,000 people, with some of the last occupation of the site estimated to have been as late as 700 BCE.

The large earthenwork complex, complete with mounds and ridgelines, stands as a testament to the complexity and sophistication of hunter-gatherers during the Archaic Period in the Lower Mississippi Valley.

THE COPPER AGE

5500 to 2000 BCE

The Copper Age, also known as the Eneolithic or Chalcolithic Era in some parts of the world, is marked by the gradual adoption of copper metal and its derived forms for toolmaking and ornamental purposes in Stone Age cultures around the world. The term Chalcolithic is derived from the Greek terms *khalkos*, meaning copper, and *lithos*, meaning stone, characterizing the use of both stone and copper tools in conjunction with each other throughout this time. The Copper Age is considered a transitional phase between the Neolithic and Bronze Age, where humans explored early forms of metallurgy while stone tools were still prevalent, and prehistory slowly became ancient history with more complex forms of symbolism and writing emerging.

In similarity to the agricultural practices of the Neolithic Revolution and toolmaking technologies of the Stone Age, the advent of copper metallurgy did not begin at one specific moment in time globally. As previously explored in the Neolithic Revolution section, the use of copper ore in North America was well established by cultures thousands of years before its widespread incorporation in Europe and the Near East.

While isolated populations had already discovered natural copper by the Neolithic Revolution, it was the widespread use of copper tools that brought many early agrarian societies out of the Stone Age. The Copper Age is often commonly divided into the Early, Middle, and Late Chalcolithic sub-periods, with exact dates varying by region globally.

The techniques for mining, smelting, and casting copper ultimately paved much of the progress towards the subsequent development of bronze and later iron metallurgy. Malachite is a mineral that was frequently extracted during this time in the Near East and Europe, which likely influenced both the use of early copper tools and the transition into the Bronze Age. It is a copper carbonate hydroxide mineral that is commonly found in oxidized layers and veins of copper deposits. Not only did it provide a means to produce cruder forms of copper, but it was prized for its color in pigments and ornaments, among other various natural dyes. Direct reduction of malachite was performed by grinding it into smaller pieces and heating it with charcoal to around 1250 °C, at which the ore decomposes and produces solidified copper oxide chunks. This copper oxide was then heated to form metallic copper in the presence of carbon atoms from wood charcoal.

Recreations of ancient malachite smelting techniques published in 2004 demonstrate this copper reduction could have been performed with relative ease and in a matter of hours with a basic controlled setup (Yee, Eddleton, and Johnson

2004). With the later addition of tin, this reinforced the atomic structure of the copper, creating a harder and more durable alloy known as bronze. The discovery and incorporation of copper metallurgy sparked significant changes in social developments, especially as the demand for copper, being a main component of bronze alloy, increased towards the Bronze Age. The transformation of toolmaking during this time would shape the efficiencies of agricultural practices, and control over copper resources along with their trade networks became a catalyst for social stratification, war, and a symbol of social status.

CHALCOLITHIC ERA

Early Chalcolithic

5500 to 4500 BCE

The Early Chalcolithic represents the primary period for the widespread emergence of early copper metallurgy and its impact on developmental changes to societies in Europe and the Near East. The gradual incorporation of metallurgy with stone and bone tools allowed more sophisticated composite tools made with copper to start appearing. Common incipient copper tools of the Early Chalcolithic include axes, chisels, knives, daggers, spear points, fish-hooks, and simple projectile points. These tools, however, were heavier, and their affixed copper became dull quickly due to its malleability. Copper ore was subjected to annealing, a process in which the metal ore was heated and slowly cooled to reduce internal stress and increase ductility, making it relatively easier to work with as a material. This provided a means to shape copper into a multitude of variations of the aforementioned early tools.

In the Near East, Egypt saw the continued rise of early settlements along the Nile River in a period before a unified kingdom known as the Predynastic Period (approximately 5000 to 3100 BCE). Early agrarian cultures such as the Tasian and Badarian are considered some of the oldest known Predynastic cultures in Upper Egypt, known for their intricate pottery, flint and limestone tools, and polished stone axes. These agricultural societies, later succeeded by the predynastic Naqada culture, are known to have raised sheep, goats, and cultivated barley and emmer wheat.

Other Late Neolithic-Early Chalcolithic civilizations include the site of Mehrgarh in the Indus Valley and the Ubaid culture of Sumer, which is known as a precursor for later development of the first proto-cuneiform writing system, appearing in Sumer's Uruk period around 3350 to 3000 BCE. In Southeastern Europe, the Baden and Cucuteni-Trypillian cultures emerged alongside the initial phase of the Early Chalcolithic. More complex ceramic traditions and advancements in architectural styles and burial rites indicated a shift in societal practices among many early Chalcolithic cultures.

The Early Chalcolithic time period should not be taken as the transitional point of widespread copper use, as this was not a sudden innovative moment in human history for every society across the globe. Rather, the gradual introduction of copper tools into the Near East, Europe, and Asia transformed ways of life dramatically during this time by reshaping trade routes and opening new avenues for agriculture with different tool technologies. Copper had already been discovered and exploited by the Old Copper Culture of North America long before this period, demonstrating that metallurgical practices could develop independently regardless of contact with societies that were already employing these toolmaking techniques.

Middle Chalcolithic

4500 to 3500 BCE

The Middle Chalcolithic period is characterized by the further development of copper metalworking, particularly in the Southern Levant, which was still supplemented by stone tool technologies. These intricately crafted copper tools included awls, spears, maces, and saws, along with many symbolic and ornamental objects that were made of copper. The Middle to Late Chalcolithic in the Southern Levant saw the rise of the Ghassulian culture from approximately 4400 to 3350 BCE in the Jordan Valley, where the type site Teleilat Ghassul is located. These pioneers of metallurgy mastered furnace techniques for smelting copper and created complex alloys using sophisticated forms of metalworking known as lost-wax casting. This was done by creating a wax model of the object, which was then encased with clay, creating a mould. The wax was subsequently melted away for copper to be poured into the mould's cavity and cooled. This allowed for intricate designs to be cast, retaining nearly identical form of the mould.

The Nahal Mishmar seasonal waterway, located in the Judean Desert of Israel, is the locus of a Ghassulian cave site that yielded 429 copper objects crafted using the lost-wax method.

Known as the Nahal Mishmar hoard, these copper artifacts, discovered in 1961, which included mace heads, vessels, crowns, and scepters, were made from a copper alloy that was rich in arsenic and antimony (Rose et al. 2023). The alloy content of these artifacts was found to be roughly 20% in atomic absorption spectrometry and electron probe microanalysis (Shalev and Northover 1993). This alloy was not used in other regions at the time in West Asia, and metallurgical investigations of petrographic characteristics revealed the ores were from Anatolia or Southern Caucasus (Goren 2008), indicating a particularly sophisticated trade network during this time. Along with their fine craftsmanship using the lost-wax technique, the rather complex metallurgy of these artifacts suggests a high level of skill among Chalcolithic artisans in the Southern Levant.

Radiocarbon dates of the Nahal Mishmar hoard revealed the occupation of the cave site to date roughly between 4000 and 3500 BCE in the Late Ghassulian period of the Southern Levant, marking a transitional phase from the Chalcolithic to Early Bronze Age in this region. The hoard was previously thought to be linked to the nearby temple of En Gedi, another significant Chalcolithic site roughly 12 kilometers away that showed signs of abandonment. This stemmed many theories about the nature of the copper artifacts found at Nahal Mishmar, and their potential to have been cultic objects from En Gedi, which were moved in secrecy to keep the hoard protected. This idea has since been challenged by critical reviews on the hoard's stratigraphic and spatial position in the cave, where new evidence suggests that this site is more closely tied to sites at the Nahal Be'er Sheva Stream based on the relationship between stratigraphy at these sites and the chronology of settlement in this region (Gosic and Gilead 2014). It also posits that the objects were not simply moved and discarded hurriedly, and were rather part of a ritualized process due to the final burial-like concealment of the hoard and the symbolic significance many of the objects represent.

As the farming and animal husbandry practices of the Neolithic spread from the Near East and Anatolia, parts of Southeastern Europe saw an increased use of copper smelting techniques, social stratification, and the establishment of extensive trade routes, specifically in the Balkans region. The sites of Rudna Glava in Serbia and Ai Bunar in Bulgaria demonstrate the use of copper for tools and ornamental purposes as copper metallurgy became more widespread by the mid-5th millennium BCE. Some of the earliest evidence of copper smelting in the Central Balkans dates back to roughly 5000 to 4500 BCE, representing the Early and Middle Chalcolithic periods. Long-distance trade networks also made it possible to exchange specialized techniques in pottery and flint-working into Central and Northern Europe, furthering the development towards the Early Bronze Age.

The Middle Chalcolithic marked a period of technological innovation in the Balkans due to these trade networks, with the widespread practice of copper smelting contributing significantly to social change and evidence of war. At the site of Tell Yunatsite in Southern Bulgaria, evidence points to a violent event occurring roughly between 4200 and 4100 BCE (Boyadzhiev 2018, 263-264). The skeletal remains of women, children, and elderly men were discovered, scattered below a sterile layer that accumulated after the site had been abandoned for more than 1,000 years. This investigation into the site is ongoing and contributes to conventional definitions on the origins of war and prehistoric combat among populations of Late Neolithic and Chalcolithic cultures.

Late Chalcolithic

3500 to 2000 BCE

Representing a transition from the Stone Age into the Early Bronze Age, the Late Chalcolithic saw the emergence of early bronze metalworking in Britain with the arrival of the Bell Beaker culture around 2450 BCE. The Bell Beaker culture is thought to have originated from the Iberian Peninsula around 2800 BCE, and they are named after their distinct bell-shaped pottery style associated with the Bronze Age transition in Europe. They are estimated to have persisted in Britain until around 1800 BCE. A general shift from using pure copper to bronze metalworking techniques saw advancements in smelting, casting, hammering, embossing, chasing, and inlaying. Bronze, being an alloy of copper and tin, was favored over its advantages in increased durability, which had a significant and long-term impact in Southeastern Europe.

Among the most widely recognized monumental architecture of Neolithic and Chalcolithic cultures in Europe, Stonehenge, located in Wiltshire, England, began construction around 3000 BCE, continuing for another 1,500 years throughout the Late Chalcolithic and into the Bronze Age. It began as a circular earthenwork ditch and bank carved out and used as a cremation cemetery for several hundred years preceding the second phase of its construction. Around 2500 BCE, the construction of the central stone began, where monumental sarsen stones and bluestones were erected on the site. Stonehenge's longevity to this day is due in part to the use of rounded tenons carved into the top of the upright sarsen stones, which fit into mortises or holes carved into the underside of the horizontal lintel stones. This created an interlocking mechanism coupled with tongue-and-groove joints between each lintel, providing stability to the structure and preventing the horizontal lintels from sliding off over thousands of years. The sarsen stones used to construct Stonehenge are a type of silcrete rock that is found throughout Southern England. Geochemical analysis of the stones using portable X-ray fluorescence (pXRF), inductively coupled plasma atomic emission spectrometry (ICP-AES), and inductively coupled plasma mass spectrometry (ICP-MS) to investigate trace element compositions and relative outcrops revealed that they were likely transported from Monkton Down, Totterdown Wood, and West Woods on the Marlborough Downs about 32 kilometers away (Ciborowski et al. 2024). Hammerstones, large amounts of sarsen and bluestone waste material, and broken stone tools were discovered in the field north of Stonehenge, indicating where the stones were shaped after transport. Its construction is attributed to many cultures spanning over hundreds of years, from the initial construction to its final modifications between 1600 and 1500 BCE, including Early Neolithic farmers and the Bell Beaker people who migrated from mainland Europe.

Perhaps one of the most widely known Late Copper Age discoveries in Europe is from the Ötzal Alps on the border between Austria and Italy. Discovered in 1991 by two German tourists hiking the Alps, Ötzi the Iceman is a naturally mummified body that was initially believed to be a deceased mountaineer on the trail. It was later confirmed through radiocarbon dating using accelerator mass spectrometry (AMS) that Ötzi existed between 3370 and 3100 BCE, placing him directly in the Chalcolithic time period of Europe (Prinoth-Fornwagner and Niklaus 1994). Besides carrying a flint knife, bow, arrows, and a quiver, Ötzi possessed a copper axe, signifying the early stages of metalworking in Europe. It is the oldest fully preserved copper axe from the Neolithic/Chalcolithic period, providing valuable insights into the technological feats and metal craftsmanship from that era (Artioli et al. 2017). The discovery of Ötzi allowed scientists to peer into a small window of the Copper Age by investigating his health, clothing, tool use, and diet, contributing to our understanding of copper metallurgy and its influence in Europe. Ötzi's body was marked with several tattoos, mainly consisting of crosses or lines, and is some of the earliest preserved evidence of this practice likely performed through a hand-poking technique (Deter-Wolf et al. 2016; Deter-Wolf et al. 2024). A recent study utilizing multiple variations of scanning electron microscopy (SEM) investigated these tattoos, revealing they were composed of soot particles along with the presence of silicate crystals, including quartz and almandine, within Ötzi's connective tissue where he had been tattooed. The crystals were likely deposited from stones of the fireplace or hearth used to make the soot (Pabst et al. 2009). Ötzi was also wearing a bearskin hat, goatskin pants, and a grass cape at the time of his death. Not only did he suffer from various ailments like Lyme disease and arthritis, Ötzi had an arrow wound in his shoulder and a fractured skull, suggesting he met a violent end, unable to escape the freezing Alps. The presence of red meat and grains in his stomach upon analysis suggests Ötzi was not only a skilled huntsman but part of an agrarian society that cultivated plant food and had already taken advantage of copper through the vast trade networks that existed throughout Europe in the Chalcolithic (Rollo et al. 2002; Nerlich et al. 2021).

In the Southern Levant, later evidence of ternary copper alloy using the lost-wax technique comes from the Late Chalcolithic site of Fazael in the Central Jordan Valley. The presence of unalloyed copper inclusions along with ternary compositions of some objects provides the earliest direct evidence of polymetallic copper metallurgy in West Asia. A large assemblage containing polymetallic copper alloy artifacts of the Late Chalcolithic period in the Southern Levant was discovered at the Fazael 2 site (Rosenberg et al. 2020). Dating to around 4000 to 3900 BCE, the items consisted of fragments of mace heads, standards, and crowns made using the lost-wax technique found in Stratum II of the large courtyard house. It is the second largest metal assemblage in the Southern Levant since the discovery of the Nahal Mishmar hoard. The 2020 publication posited that the presence of crucible and lost-wax cast alloy fragments may indicate a former lost-wax casting site present in Fazael. However, Rosenberg et al. point out that the evidence of a furnace has yet to be uncovered in Fazael, noting the unusually large number of copper alloy artifacts could instead indicate that the site was used for the recycling of polymetallic copper alloys cast from the lost-wax technique. The addition of flint artifacts in this assemblage may indicate the early use of Canaanean blades, often associated with the Early Bronze Age, which challenges the understanding of Copper Age timelines in the Southern Levant. It also highlights the onset of Bronze Age techniques and represents a late stage of the Late Chalcolithic in the region.

The transition from copper use into the Early Bronze Age traditions and practices throughout Europe occurred between the late 5th and the late 3rd millennium BCE. In the Near East, it began in the late 5th millennium BCE and lasted until the 4th millennium BCE before the rise of the Bronze Age. As tin became more frequently used in copper alloys to create bronze tools and ornaments favored for durability, many earlier polymetallic copper alloys were replaced, marking a significant transition in cultures of the Near East and Europe.

THE BRONZE AGE

3300 to 1200 BCE

The Bronze Age marked a significant transformation in metallurgy as tin became more prevalently used for polymetallic copper alloys. The use of tin, copper, lead, and arsenic to create bronze tools was favored over earlier alloys of the Chalcolithic for their durability and hardness over copper. Bronze tools would reshape systems of trade, agricultural practices, and hierarchical social systems over the next several thousand years, laying the foundations for civilizations in Ancient History. Distinct parts of the world entered their own Bronze Ages at different times as cultures began to incorporate bronze alloys in toolmaking and ornamental purposes.

In places like Greece and China, the Bronze Age began before 3000 BCE, whereas in Britain it did not start until about 2500 BCE. The beginning of the Bronze Age sometimes includes the end of the Chalcolithic Era in sites like Tell Yunatsite and Uruk, where growth during the Late Chalcolithic period continued into the Early Bronze Age. These overlapping periods of coeval tool use demonstrate longer periods of societal stability and that the transition to incorporating bronze technologies was gradual instead of an immediate replacement.

By around 3500 BCE, the use of copper to cast small tools and weapons was rapidly developing in the Near East and Southeast Europe, becoming a leading factor for urbanization in and around Mesopotamia. Around 3000 BCE, the use of copper was well known in the Middle East, had extended west into the Mediterranean area, and was beginning to infiltrate the Neolithic cultures of Northern and Central Europe. Although the Chalcolithic is commonly thought of as part of the Bronze Age, true bronze alloy made of copper and tin was used only rarely at first, distinguishing these periods of extensive adoption between metals.

Between 2000 and 1000 BCE, the use of true bronze alloy became more widespread. This age also saw a refinement of the wheel and the invention of the ox-drawn plow. Metalworking techniques from the Bronze Age include embossing (also known as repousse), chasing, engraving, inlaying, enameling, and gilding, which became hallmarks of bronze metallurgy throughout Europe, the Near East, Sub-Saharan and North Africa, and Southern Asia, among other isolated instances of early bronze use in Chalcolithic cultures.

Early Bronze Age

3300 to 2100 BCE

The start of the Bronze Age is often largely attributed to beginning in Mesopotamia and the larger region of the Fertile Crescent. This was marked by the rise of urban centers and the further development of Sumerian civilization following the Ubaid period, which lasted from 6500 to 4100 BCE, where settled agricultural life first emerged in the region. Sumer was not a single unified nation, and instead was a collection of independent city-states, each with its own temple known as a ziggurat, ruler, and territory, often competing or warring with each other. The Early Bronze Age brought the initial development of bronze metallurgy and its impact on tool and weapon production in this region of developing city-states between the Tigris-Euphrates river system. The use of copper was still predominant during this time in certain parts of the world like Mesopotamia, the Middle East, Mediterranean civilizations (such as the Minoan occupation of Crete, the Mycenaeans on Greece, and the Cycladic culture on the Cyclades Islands), Egypt, some Neolithic cultures of Europe, and Anatolia (present day Turkey) as tin gradually became more commonly used in copper metallurgy to create bronze alloys. While some of the earliest bronze artifacts are found in Mesopotamia, there were no substantial tin deposits in the region during the Bronze Age, suggesting advanced trade networks from Anatolia, Iran, and Afghanistan, where tin ore was accessible.

Urbanized civilizations such as Uruk in the southern alluvial plains of Mesopotamia and the Indus Valley Civilization (IVC) in the northwestern region of South Asia continued to develop from Chalcolithic Era technologies and became pivotal societies of the Bronze Age. The rapid expansion of bronze metallurgy throughout these regions brought with it the rise of city-states. As one of the most important cities in ancient Sumer, Uruk rose to prominence between the Tigris and Euphrates rivers following the Ubaid culture in the Chalcolithic Era around 4000 BCE, which gradually developed into a powerful Bronze Age city-state and flourished for centuries after its construction. At Uruk's peak, the city was estimated to have had a population capacity of 40,000-80,000 people. It is often referred to as the "First City" for its monumental architecture and social hierarchy system, which created an epicenter of urban life in Mesopotamia. By the Late Uruk period in Sumer, a proto-cuneiform writing system of pictographs and symbols impressed into clay tablets began to appear around 3350 to 3000 BCE, the earliest and first known writing system in human history. In 1853, archaeologists discovered a collection of Akkadian clay tablets at the ruins of the Royal Library of Ashurbanipal in Nineveh, one of which contained a version of the Epic of Gilgamesh. Written in Cuneiform sometime around 2100 BCE, the epic describes the legendary king Gilgamesh of Uruk, who rose to power sometime between 2800 and 2500 BCE and is considered the world's oldest poem, predating the works of Homer by roughly 1500 years. It provided a fundamental insight into religious beliefs and the various human experiences of Bronze Age societies in Mesopotamia, which later influenced many forms of literature and art over the next several centuries.

By around 3100 BCE, Egypt was unified into the Old Kingdom under King Narmer, also known as Menes, which marked the beginning of the First Dynasty in Egypt. A ceremonial slate tablet depicting the unification, known as the Palette of King Narmer, remains one of the earliest known instances of Egyptian hieroglyphics. Royal tombs initially began as rectangular structures with flat roofs known as mastabas. The construction of the first pyramid in Egypt began centuries later in the 27th century BCE during the reign of King Djoser in the Third Dynasty, with the Step Pyramid at Saqqara. The first true pyramid with smooth sides, however, was constructed around 2600 BCE at Dahshur during the rule of Sneferu, the first king of the Fourth Dynasty. It is known as the Bent Pyramid, due to its angle of slope being changed during construction to prevent collapse, resulting in its bent shape. Sneferu is often attributed to being responsible for completing the Meidum Pyramid, which was started by his supposed father, Huni, and commissioning the later construction of the Red Pyramid and Seila Pyramid, which is a lesser-known structure. Most famously known is the Great Pyramid of Giza, commissioned by Pharaoh Khufu, second king of the Fourth Dynasty, to be built around 2580 to 2560 BCE as his final resting place. The two other pyramids of the Giza pyramid complex were built following this for Khafre, the fourth king, and Menkaure, the fifth king of the Fourth Dynasty.

In Mesopotamia around 2700 BCE, the Sumerian city-state of Kish engaged in what is considered to be the first recorded evidence of military conflict. This event took place with the Kingdom of Elam, located in present-day Iran, and is highlighted in a Sumerian text known as the Sumerian King List. This text is preserved on cuneiform inscribed clay tablets and a clay prism known as the Weld-Blundell Prism, which lists the kings of Sumer, the city-states in which they ruled, and the length of their reign. It blended historical and embellished mythological accounts with some rulers unrealistically noted to have reigned for tens of thousands of years. Among these listings, it describes King Enmebaragesi of Kish leading the Sumerians to victory after seizing the weapons of Elam, suggesting a disarmament of the kingdom. This account is one of the earliest recorded wars in human history.

In present-day Pakistan and northwest India, the Indus Valley Civilization (IVC), also known as the Harappan Civilization, began to take shape around the Late Uruk period by 3300

BCE. Urbanization and extensive trade networks with Mesopotamia led to the later development of two large city-states within the Indus Valley known as Harappa and Mohenjo-Daro around 2600 to 1900 BCE. These two cities featured sophisticated urban planning with defined streets and standardized brick sizes. The development of street drainage systems and well-water access further bolstered sanitation and public health by reducing the spread of waterborne diseases and promoting the growth of these two cities (Sharma 2020). Other notable settlements in the Indus Valley included Kalibangan, Dholavira, Rakhigarhi, Rupnagar, and Lothal, all of which adopted the Harappan script writing system, which developed in the Indus Valley around 2800 to 2600 BCE. By 2334 BCE, the Akkadians rose to prominence after the conquest of Sumer, which united Mesopotamia under one ruler led by Sargon the Great. He later expanded his empire further by conquering the Kingdom of Elam, parts of Syria, and Anatolia. Sargon the Great is often considered to be the world's first ruler of a multinational empire following these series of conquests.

Middle Bronze Age

2100 to 1550 BCE

The Middle Bronze Age was a period of expansion and the further development of larger empires, with civilizations creating complex writing systems after the introduction of cuneiform in the Early Bronze Age. Urban planning became more advanced as tools made of true bronze alloy circulated more prominently in Europe, the Near East, and East Asia. In Europe, the Middle Bronze Age period was characterized by the spread of early Scottish settlement traditions further south, and the development of ring-ditch house settlements. Stonehenge was also in stages of continued development with subsequent modifications and burials added until around 1500 BCE. The Mycenaeans represented the first advanced and distinctively Greek civilization on mainland Greece during this period of the Bronze Age, spanning from about 1750 to 1050 BCE. Adopting the culture and technology of the Early Bronze Age Minoan civilization on Crete, the Mycenaeans became masters of bronze metallurgy. Extensive trade routes were established between the Mycenaeans and Minoans, which enabled the exchange of pottery, metallurgical techniques, and agricultural technology. While Minoans were known for creating a wide range of bronze tools to improve their agrarian foundation, the Mycenaeans were a militaristic society focused on using bronze alloys for weapon-making and armor. The Minoans are credited with creating the writing system known as Linear A script, with Linear B script later becoming an adaptation of Linear A created and used by the Mycenaean Greeks as an early form of the Greek language. In a 2017 publication that explored the genetic origins of these two ancient cultures, it is noted that the Minoans were Europe's first literate civilization; "However, the Linear A syllabic ideographic and Cretan hieroglyphic scripts used by this culture remain undeciphered, obscuring its origins" (Lazaridis et al. 2017). The Minoans, who existed on the island of Crete from about 2700 BCE, shared a significant period of overlapping history with the Mycenaeans before being conquered by them in 1450 BCE.

The Middle Bronze Age also brought a renewal of urban settlement in Mesopotamia and the Near East, with the construction of fortified cities containing complex societies that moved away from technologies of the Stone Age. The Akkadians ruled until 2154 BCE, when they were conquered by the Gutians, a nomadic people of the Zagros mountain range who launched a full-scale invasion leading to the empire's collapse. The Gutians controlled Sumer for approximately a century until they were driven out in a successful rebellion by Utu-Hegal of Uruk, the first native king since Akkadian rule over Sumer. Following the fall of Akkadian rule, the empires of Babylonia and Assyria began to take shape after the Amorites conquered

Mesopotamia between 2000 and 1900 BCE. The first city of the Assyrians known as Assur, was founded in the 3rd millennium BCE and later became the capital of the empire. Babylon became the capital of the Babylonians under the Amorite king Hammurabi, widely known today for creating the Code of Hammurabi. Written in Akkadian around 1754 BCE, it is one of the first recorded complete codes of laws that describes principles of retributive justice, where the punishment should be equivalent to the crime.

Following the decline of the Old Kingdom in Ancient Egypt, the First Intermediate Period marked a time of political instability with the centralized powers of Pharaohs becoming increasingly illegitimate to the new regional powers of Upper and Lower Egypt. This period lasted from about 2181 to 2055 BCE until the Middle Kingdom was established under Mentuhotep II, who reunified Egypt. This became a time of renewed stability in the region with advancements in agriculture until another decline in Egypt during the Second Intermediate Period from about 1782 to 1550 BCE. This was later followed by Egypt's Golden Age, the New Kingdom, which brought with it powerful pharaohs such as Ramesses II, Tutankhamun, Akhenaten, and Hatshepsut. The construction of the temples at Luxor, Karnak, and the Valley of the Kings took place during this period. It is often considered the most prosperous time in Egypt with advancements in bronze weaponry, which bolstered military conquests of Nubia and the Near East, significantly expanding Egypt's influence in the region.

Late Bronze Age

1550 to 1200 BCE

The Late Bronze Age is known for the subsequent Late Bronze Age Collapse, which was the dramatic decline of many Near Eastern and Mediterranean civilizations along with the transformation of Egypt. Preceding the collapse, bronze metallurgy of weapons such as leaf-bladed swords, spearheads, and the construction of chariots dramatically shaped Late Bronze Age cultures and militaristic societies. These advancements had a significant impact on warfare, influencing both tactics and social structures of armies. This period witnessed the continued advancements of powerful civilizations with interconnected trade networks and complex societies including the Mycenaean occupation of Greece, the New Kingdom of Egypt, Babylon, Assyria, the Indo-European Hittite Empire, Cyprus, and the cities of Canaan which encompassed present-day Palestine, Jordan, Lebanon, Israel and Syria (Linkov et al. 2024). By the end of the Late Bronze Age, these civilizations built monumental architecture, used advanced bronze metallurgy, and developed complex literacy.

In China, the Yin-Shang Dynasty rose to prominence around 1600 BCE, with bronze metallurgy becoming increasingly practiced. Earlier evidence of bronze use in China dates to around 3000 BCE with the discovery of a bronze knife in the Gansu Province associated with the Majiayao culture, created using section-mold casting. The bronze knife marking the beginning of the Bronze Age in China has been the subject of debate, as it is considered an isolated instance of bronze with no associated smelting site. Furthermore, it was discovered in the western part of the Loess Plateau, which is covered by a thick Quaternary loess, limiting access to hard ore containing rock (W. Sun et al. 2016). There has also been no evidence reported of a substantial ore deposit near the site, either, suggesting it was imported or traded. However, the development of elaborate bronze vessels and ritual containers in China reached its peak during the Late Bronze Age.

The Canaanite port city of Ugarit, located in present-day Syria, reached its height during the Late Bronze Age. While being settled much earlier during the Neolithic Era, possibly by as early as the 7th millennium BCE, Ugarit established extensive trade routes by around 1500 BCE, ultimately creating strong ties with Egypt, Cyprus, and the Mediterranean throughout the Late Bronze Age period. Egypt was still in its New Kingdom period and began vying for control of the Levant, leading to the Battle of Kadesh in 1274 BCE against the Hittite Empire. The Hittite Empire rose to prominence likely after arriving in Anatolia around 2000 BCE, and later established its capital, Hattusa, around 1600 BCE. After consolidating power and expanding into Syria during their Old Kingdom period, the New Kingdom period saw the rise of military campaigns in the Late Bronze Age. The Battle of Kadesh was a noteworthy skirmish between Egypt and the Hittite Empire, resulting in the largest chariot battle in ancient history, with between 5,000 and 6,000 chariots deployed in the clash. The battle resulted in a stalemate, with both sides eventually leading to a peace treaty roughly 15 years later, which is considered the earliest known written international peace treaty.

Late Bronze Age Collapse

1200 to 1150 BCE

The Late Bronze Age Collapse is a well-known period of drastic societal disruption and decline that occurred in the Eastern Mediterranean and the Near East, including regions like Egypt and Anatolia, Southeastern Europe, and the Aegean region. Around 1200 BCE, many societies of the Bronze Age began to experience a rapid decline, with various causes being the subject of contributing factors to this widespread collapse of empires and civilizations. In 1177 BCE, a group of seafaring people known colloquially as Sea Peoples invaded Egypt, becoming a catalyst for a period of decline in the region despite victory over the invaders (Cline 2014). With many surrounding civilizations in Mesopotamia, the Mediterranean, and other parts of the Near East also experiencing this decline, the Late Bronze Age Collapse is attributed to drought, climate change, natural disasters such as earthquakes, and internal political strife. What followed was centuries of disorganization, plunging many parts of the Levant, the Near East, Southwest Asia, and the Eastern Mediterranean into periods of political fragmentation, a decrease in monumental art and architecture, disruption of trade routes, and population decline.

While the causes of the sudden destruction of major cities, the decline of complex civilizations, and the subsequent rise of new powers are debated, factors like climate change, invasion by Sea Peoples, and social unrest are commonly cited. Major civilizations like the Mycenaeans and Minoans were absorbed into other cultures and more decentralized settlements, marking the beginning of the Greek Dark Ages. The Hittites effectively disappeared from much of the historical record, and many important cities were destroyed, including Mycenae and Ugarit. The Kingdom of Elam and Phoenician city-states continued to survive the collapse along with Assyria in northern Mesopotamia, though it lost territory to the Kingdom of Elam in its southern region.

The New Kingdom of Egypt also survived this devastating decline. Although this would cause Egypt to experience periods of disorganization, it maintained power and influence throughout the collapse until it entered the Third Intermediate Period in 1070 BCE. As iron ores became more readily available due to a shortage of tin, a key ingredient in bronze, iron smelting techniques became more proficient. These surviving civilizations of the Bronze Age would eventually set the stage for the transition into the Iron Age. With many civilizations and empires previously relying on trade networks for copper and tin ores, the instability and disruption of these trade networks during the collapse may have contributed to the scarcity of these materials. This scarcity, along with the cultural and societal disruption of skilled labor, technologically atrophied some bronze metallurgy techniques until their gradual replacement by iron smelting at the beginning of the Iron Age.

THE IRON AGE

1200 to 550 BCE

The date at which iron replaced bronze almost completely in tool and weapon making varies widely geographically. The Late Bronze Age Collapse became a catalyst for the transition to iron smelting by many civilizations, as mentioned in the previous section. Ancient Egypt fell into a period of decline and fragmentation, marking the Third Intermediate Period (1078 to 664 BCE). This recession made it difficult for Egyptian society to fully incorporate ironmaking technologies until much later, around the 6th century BCE. While there is earlier evidence of iron use dating to the Naqada II Period in Egypt (ca. 3200 to 3000 BCE) for simple ornamental purposes, its widespread adoption did not occur until near the Late Period of Egypt (664 to 332 BCE), when Egypt fell under foreign rule. As new metallurgical practices began to replace bronze smelting techniques in Egypt, particularly with iron, and the allure of immense gold resources, ambitious foreign powers such as the Assyrians, Persians, and Macedonians all began vying for control of Egypt and its resources.

The use of iron began in the Middle East and Southeastern Europe around 1200 BCE, while in China, this technology did not become substantial until around 600 BCE during the

Zhou Dynasty. As the Iron Age neared proto-historic and historical periods, once the knowledge and use of iron metallurgy became widespread, geographic subperiods within the Iron Age emerged. Each subperiod has specific regional variations and alternative classifications. In the Near East and Europe, the Iron Age is divided into two primary subperiods: Iron Age I (1200 to 900 BCE) and Iron Age II (900 to 550 BCE). Iron Age II is commonly studied in further divisions, namely Iron Age IIA, IIB, and IIC.

In the Americas, the timing of the Bronze-Iron Age transition corresponds to the Woodland Period in North America, and the Formative Period in both Meso and South America, which began around 1500 to 1000 BCE. It is important to remember that cultures in the Americas did not undergo a Bronze or Iron Age like that of Europe and the Near East, and instead developed distinct and sophisticated forms of copper, bronze, and gold metallurgy independently. Woodland cultures of North America are known for intricate pottery styles, diverse agricultural practices, the construction of burial mounds, and sedentary village lifestyles. These included the Adena, Hopewell, Mississippian, and Plains Woodland cultures, among other societies that gradually saw the rise of associated language families. These include Iroquoian, Algonquian, and Siouan, representing the groups that emerged within or became influenced by these Woodland cultures.

Cultures such as the Olmec of Mesoamerica and the Chavin, Nazca, Moche, and Valdivia of South America transformed later Andean societies throughout the Formative Period. It was a period of increased construction of megalithic monuments with intricate art styles and the development of complex hierarchical societies. Along the temperate rainforest basins in Ecuador, the Valdivia culture (ca. 3500 to 1500 BCE) thrived near the coastline and was known for incorporating intricate motifs and symbolism represented in their pottery styles. Being one of the oldest cultures in the Americas with a settled lifestyle, their early advancements in pottery are evident by the distinct style and technique to turn practical pieces into more elaborate and symbolic works. One of the hallmarks of Valdivian culture was their creation of pottery figurines. These are characterized by their intricately detailed hairstyles and may have been representations of actual people or used as votive objects. However, the imposed characteristics of fertility and sexuality on these prehistoric figurines by Western thinkers have created controversy about their true purpose, when their meanings are truly unknown. Attempts to impose modern associations onto prehistoric objects such as these undermine the individual complexities of past cultures.

While Valdivian pottery is most often associated with their iconic figurines, their refined pottery-making techniques produced elaborate and distinctive pieces. Frequent styles of

Valdivia pottery involved burnishing techniques to create a polished surface. By rubbing a smooth object, such as a stone or bone, over the surface of partially dried clay, the particles of clay become compacted, forming a smooth and sometimes glossy surface. Other styles incorporated the use of dark red slips, a thin clay coating applied before firing, which created a dichromatic effect on the vessel, where both red slip and natural clay color are present after firing.

Valdivian settlements were typically organized in circular patterns around a central plaza. They lived a semi-agrarian, sedentary lifestyle, cultivating crops such as maize, squash, and beans, supported by fishing and hunting. The Valdivia are known to have engaged in trade along the Pacific coastline of South America using rafts with sails to establish trade routes with other groups surrounding the Andes and Amazonian regions. They played a crucial role in the development of later Andean civilizations by influencing pottery, agriculture, and trade practices.

IRON AGE SUBPERIODS

Iron Age I

1200 to 900 BCE

In the earlier phases of the Iron Age throughout the Mediterranean, Near East, and Europe, iron metallurgy began to spread gradually out of civilizations in Anatolia and the Hittite Empire. The nucleus of the Iron Age is often thought to have begun in Anatolia, the Mediterranean, and parts of the Near East following the Late Bronze Age Collapse of the Hittite Empire and the Mycenaeans around 1200 BCE, ushering in the Greek Dark Ages. This was a period of societal regression, atrophy of the Linear B writing system and literacy, the disruption of trade networks, and a decline in population, cultural movements, and economic activity in the Mediterranean.

At the archaeological site of Kaniş Karum, located in the Kayseri Province of present-day Turkey, also known as Kültepe-Kaniş, approximately 23,500 clay tablets were unearthed, written in Old Assyrian cuneiform. These tablets describe the commercial trade of iron and copper along with treaty information between Assyrian colonies and colonies of the Hittite Empire (M. Kaya and G. Kaya 2024). They are the earliest written documents of ancient Anatolia, and also highlight the social complexity of the Hittite and Assyrian empires before this major collapse. The clay tablets shed light on the personal lives, legal matters, and a woman's right to inherit their husband's estate within the Old Assyrian Empire.

It is important to note that the diffusion of ironworking did not always happen at the same time, nor through means of a simple transfer of knowledge. Different regions entered their own Iron Age at different times, adopting and adapting the technology based on each society's specific needs or goals. While the Hittite Empire played a crucial role in the development of iron metallurgy before the Late Bronze Age Collapse, the spread of this technology was a complex, multifaceted process, and it is highly likely that other cultures developed iron technology independently. The collapse of the Hittite Empire around 1180 BCE left trade networks through Anatolia disintegrated, necessitating skilled blacksmiths to migrate into surrounding regions. Neighboring civilizations such as the Assyrians in the Southern Levant began to adopt iron-making technologies around this same time, marking a significant shift in expansion and military dominance for the Neo-Assyrian Empire following a period of decline in the Middle Assyrian Empire.

Over in the region of present-day India, the widespread use of iron tools and weapons significantly impacted various aspects of early Indian societies. The Iron Age I is also associated with the Vedic Period in India (ca. 1500 to 500 BCE), following the decline of the Indus Valley Civilization, reflected in pottery traditions. This period in India saw the rise of new pottery cultures, including Painted Grey Ware (Ghaggar-Hakra valley and the Gangetic Plains) and Northern Black Polished Ware (Ayodhya and Sringaverapura). These pottery styles are wheel-thrown and characterized by their grey color shades, black geometric designs, and retaining a thin vessel wall. Moreover, the Painted Grey Ware pottery style is a significant indicator of the Northern Indian Iron Age. The Painted Grey Ware culture is thought to have been succeeded by the Northern Black Polished Ware culture, which is often associated with the rise of Mahajanapada states and the Magadha Empire. This early stage of the Iron Age in India saw the beginnings of powerful states as well as the development of new religions and philosophical movements, including Hinduism, Buddhism, and Jainism.

By this time, Egypt had entered its Third Intermediate Period, marking a significant societal fragmentation and the delay of widespread iron-making over bronze smelting. Egypt relied on imported iron from Asia and the Middle East, particularly from the Hittite Empire, for tools and weapons due to a lack of local resources and the necessary materials to smelt iron. Following the fall of the Hittites, the network of routes facilitating trade with Anatolia and the surrounding regions disappeared almost entirely, causing the supply chains of iron ore to bottleneck. As a result, Egypt primarily used iron for making trinkets and jewelry during this time, instead of focusing on tool or weapon making. These were made with hematite iron and sienna, a type of earth pigment derived from ferric oxide found in clays. Egypt would later undergo a series of conquests in the Iron Age II period, resulting in a final conquest by the Macedonians under Alexander the Great in 332 BCE, which dismantled Persian control of Egypt. This event ultimately shaped the beginning of the Classical Antiquity Period and the Hellenistic Age, and is marked by the spread of Greek culture throughout the Mediterranean, Near East, Europe, Levant, and even parts of Southeast Asia.

Iron Age II

900 to 550 BCE

The second half of the Iron Age began with a more widespread use of ironmaking, driven by the necessity for advancements in metallurgy. This played a crucial role in the revitalization and reshaping of many societies following the Late Bronze Age Collapse. Trade patterns shifted in Europe, connecting South and West-Central Europe with the Mediterranean, leading to a more direct spread of Greek culture following the end of the Greek Dark Ages. This transitioned Greece into the Archaic Period around 800 BCE, which saw the emergence of early Greek art and philosophy, such as the works of Homer, and the establishment of the city-states Athens and Sparta in the southern region of Greece. The Etruscan civilization flourished in present-day Tuscany, Italy, from around the 8th century BCE until their culture was gradually absorbed by the Roman Republic. After Rome was established in 753 BCE, the Roman Republic was founded in 509 BCE following the overthrow of the last Etruscan king who had ruled Rome. In Central Europe, the Hallstatt culture emerged from the Late Bronze Age Urnfield traditions, a crucial development for proto-Celtic culture. Eastern Europe and the Caucasus region saw the rise of the Koban culture (ca. 1200 to 350 BCE) and the Chernogorovka-Novocherkassk complex (ca. 900 to 650 BCE) near the Pontic Steppe and Caucasus Mountains. The Koban culture of northern and central Caucasus developed agrarian settlements and was known for advanced metallurgy, especially in bronze work preceding the Iron Age. In some parts of Scandinavia and Northern Europe, the Iron Age continued until 800 to 900 CE, during the transition into the Viking Age.

Around 550 BCE, the Achaemenid Persian Empire began rising to prominence under Cyrus the Great, centered in present-day Iran. The detailed and reliable records and historical accounts of the Achaemenid Empire brought an end to the proto-historical Iron Age in the Ancient Near East. The two subsequent Persian invasions of Greece during the Greco-Persian Wars from 499 to 449 BCE later ended the Archaic Period in Greece. Specifically, the Greek victories at the battles of Thermopylae and Salamis during the second invasion marked key transitional events of Greece entering the Classical Period. This period is often called the "Golden Age of Greece" and is commonly associated with the rise of Western philosophy from influential Greek philosophers and scholars, namely Socrates (470-399 BCE), his student Plato (428-348 BCE), and Plato's student Aristotle (384-322 BCE).

Fueled by the aftermath of the Persian invasions in Greece, Sparta and Athens engaged in their own conflict known as the Peloponnesian War from 431 to 404 BCE. This saw a dramatic reshaping in the political system of Greece following the

Spartan victory over Athens. Athens' democracy was replaced by an oligarchy, significantly weakening the political landscape in Greece. In 359 BCE, the Macedonian Empire rose to power under Philip II, who later conquered Sparta and Athens in 338 BCE after the Battle of Chaeronea. This victory established Macedonian hegemony over Greece, followed by Sparta's resistance and the eventual assassination of Philip II by one of his bodyguards, Pausanias of Orestis, in 336 BCE. The Macedonians later solidified their control over Greece under Alexander the Great, the son of Phillip II, by suppressing the rejection of Macedonian authority in the city of Thebes in 335 BCE. In 334 BCE, Alexander the Great began his series of conquests and a campaign against Persia that would define his reign.

As the Vedic Period continued in India, the growth of power and influence in the initial tribal settlements, known as Janapadas, led to the establishment of larger organized states known as Mahajanapadas. By the end of the Vedic Period around the 6th century BCE, sixteen major Mahajanapada kingdoms existed in ancient India. This marked the beginning of the Mahajanapada Period, with one of the kingdoms known as Magadha eventually becoming the dominant power and consolidating its rule over northern India. The Magadha Empire led to increased urbanization and paved the way for the establishment of the Nanda Dynasty around 345 BCE. From 327 to 325 BCE, Alexander the Great led an incursion into India, exposing the weaknesses of existing kingdoms and creating a power vacuum that Chandra Gupta Maurya exploited following the withdrawal of Alexander's forces. The fragmentation of these kingdoms, coupled with the power vacuum Alexander the Great had left behind, allowed Chandra Gupta to reunify much of the Indian subcontinent under the Mauryan Empire, which overthrew the Nanda Dynasty in 322 BCE. This event further unified India and expanded the region, marking the end of the Iron Age in Ancient India and the beginning of the Classical Antiquity Period in the region between the 3rd and 2nd century BCE.

Egypt transitioned into another period of destabilization under foreign rule with the beginning of the Late Period in 664 BCE. This began with the conquest by the Assyrians, in which they installed a puppet king, Psamtik I. Psamtik I initially served under Assyrian supremacy, until he later strategically reformed Egypt's government to leverage the decline of Assyrian influence and restore native Egyptian rule. By 525 BCE, Egypt fell into Persian rule, starting with the conquest under Cambyses II, the son of Cyrus the Great. This would ultimately lead to the later conquest by the Macedonians under Alexander the Great in 332 BCE, in which his army dismantled Persian control of Egypt and established the city of Alexandria. This conquest ushered in the early stages of the Classical Antiquity Period and the Hellenistic Age influence in Egyptian society. In the aftermath of Alexander the Great's death in 323 BCE following the invasion of India, his empire was divided among the Macedonian generals, with Ptolemy I Soter establishing the Ptolemaic

Dynasty in Egypt. This ultimately brought Greece and Egypt into the Hellenistic Age of Classical Antiquity, characterized by the spread of Greek culture through the Mediterranean and Near East following the conquests.

THE HISTORICAL PERIODS

Classical Antiquity

8th century BCE to 6th century CE

Post-Classical Period (Middle Ages)

5th century to late 15th century CE

Early Modern Period

1453 CE to 1800 CE

Late Modern Period

1800 to 1945

Contemporary History

1945 to Present

ARCHAEOLOGICAL PERIODS

North America

Lithic/Paleoindian (pre 8000 BCE)

Archaic (8000 BCE to 1000 BCE)

Woodland (1000 BCE to 1000 CE)

Mississippian (800 CE to 1600)

Mesoamerica

Lithic/Paleoindian (pre 8000 BCE)

Archaic (8000 BCE to 1000 BCE)

Formative (1000 BCE to 250 CE)

Classic (250 CE to 900 CE)

Post-Classic (900 CE to 1515)

South America

Lithic/Paleoindian (pre 8000 BCE)

Archaic (8200 BCE to 1000 BCE)

Formative (1000 BCE to 500 CE)

Early Intermediate Period (200 BCE to 600 CE)

Classic (500 CE to 1200)

Late Intermediate Period (1000 CE to 1476)

Post-Classic (1200 CE to 1900)

Northern Europe

Nordic Stone Age/Nordic Bronze Age (1700 BCE to 500 BCE)

Pre-Roman Iron Age (500 BCE to 1 BCE)

Roman Iron Age (1 CE to 400 CE)

Germanic Iron Age (400 CE to 800 CE)

Viking Age (800 CE to 1066 CE)

Medieval Period (1066 CE to 1500)

Post-Medieval Period (1500 to 1800)

Western Europe

Paleolithic (pre 8800 BCE)

Mesolithic (8800 BCE to 4900 BCE)

Neolithic (4900 BCE to 2000 BCE)

Bronze Age (2000 BCE to 800 BCE)

Iron Age (late 11th century BCE to 1 BCE)

Roman (56 BCE to 400 CE)

Early Medieval Period (400 CE to 800 CE)

Medieval Period (800 CE to 1500)

Post-Medieval Period (1500 to 1800)

Southeastern Europe

Paleolithic (1.5 million years ago to 8000 BCE)

Epipaleolithic (20,000 years ago to 8000 BCE)

Neolithic (7000 BCE to 3000 BCE)

Chalcolithic (5300 BCE to 3500 BCE)

Bronze Age (3300 BCE to 700 BCE)

Iron Age (1200 BCE to 100 BCE)

Hellenistic Period (323 BCE to 30 BCE)

Roman Period (200 BCE to 500 CE)

Byzantine Period (330 CE to 1453)

Ottoman Empire (1354 to 1804)

North Africa

Paleolithic (2.4 million to 25,000 years ago)

Epipaleolithic (25,000 years ago to 8000 BCE)

Neolithic (8000 BCE to 3000 BCE)

Bronze Age (3300 BCE to 1200 BCE)

Iron Age (750 BCE to 1000 CE)

Roman Period (146 BCE to 395 CE)

Sub-Saharan Africa

Earlier Stone Age (3.3 million to 280,000 years ago)

Middle Stone Age (300,000 to 25,000 years ago)

Later Stone Age (30,000 to 5,000 years ago)

Neolithic (4000 BCE to 3000 BCE)

Bronze Age (3500 BCE to 600 BCE)

Iron Age (550 BCE to 700 CE)

Classic Middle Ages (700 CE to 1700)

Levantine

Stone Age (2 million years ago to 3300 BCE)

Chalcolithic (4500 BCE to 3300 BCE)

Bronze Age (3300 BCE to 1200 BCE)

Iron Age (1200 BCE to 586 BCE)

Historical Periods (586 BCE to Present)

Southern Asia

Paleolithic (53,000 years ago to 10,000 BCE)

Mesolithic (10,000 BCE to 6500 BCE)

Neolithic (6500 BCE to 2000 BCE)

Chalcolithic (4000 BCE to 2000 BCE)

Bronze Age (3100 BCE to 1100 BCE)

Iron Age (1100 BCE to 500 BCE)

Proto-Historical/Vedic Period (1500 BCE to 500 BCE)

Historical Periods (after 500 BCE)

Korea

Paleolithic (40,000 years ago to 8000 BCE)

Jeulmun Pottery Period (8000 BCE to 1500 BCE)

Mumun Pottery Period (1500 BCE to 300 BCE)

Protohistoric Period (300 BCE to 400 CE)

Three Kingdoms of Korea (300 CE to 668 CE)

Japan

Paleolithic (35,000 years ago to 10,000 BCE)

Jōmon Period (10,000 BCE to 300 BCE)

Yayoi Period (300 BCE to 250 CE)

Yamato Period (250 CE to 710 CE)

China

Paleolithic (1.7 million years ago to 8000 BCE)

Neolithic (10,000 BCE to 2100 BCE)

Ancient China (2100 BCE to 221 BCE)

Imperial Period (221 BCE to 1911 CE)

Republic of China Era (1911 to 1949)

Modern Period (1949 to Present)

Australia

Indigenous Period (60,000 BCE to 1606 CE)

European-Contact Pre-Settlement Period (1606 to 1788)

Settlement/Pre-Industrial Period (1788 to 1820)

Industrial/Modern (1820s to Present)

New Zealand

Archaic Period (1000 CE to 1650)

Classical Period (1350 to 1800)

Eastern South Island Classical Periods (1650 to 1800)

GEOARCHAEOLOGY

To be what we are, and to become what we are capable of becoming, is the only end of life.

- Robert Louis Stevenson

GEOARCHAEOLOGY IS DEFINED AS A SPECIALIZED interdisciplinary subfield of archaeological inquiry and study that applies methods from the earth sciences, chemistry, biology, and physics to answer archaeological questions about past human settlement activity, depositional events, and the causes of major paleoenvironment changes. Geoarchaeology encompasses all archaeological investigations, as cultural material and remains are deposited in geological situations. It is the discipline used to diagnose the long-term conditions that affect buried or submerged cultural material in the past and present. For prehistoric contexts, geoarchaeology is used as a method of investigating the past through the sediment and soil, which acts as a record before the advent of written text and historical documentation. For this reason, it is crucial for the practicing archaeologist to maintain a working knowledge and habituated understanding of how Earth's geological processes influenced the formation of the archaeological record. It is also important to understand the relationship between archaeological material and the depositional circumstances from which they were uncovered.

As past human settlements changed over time, the materials left behind became subject to geological processes and events that impacted the spatial distribution and preservation of those remains, evident in layers of stratigraphy. Pedology, or the study of Earth's surface soil, is another multidisciplinary field from which geoarchaeology derives techniques. This includes soil classification, distinguishing between buried paleosols (ancient land surfaces; pedostratigraphy) and sedimentary formations (lithostratigraphy), and determining the properties of soil horizons. By analyzing soils, sediments, and microfauna in archaeological sites, geoarchaeologists can associate post-depositional processes with site formation history. In the field, this is typically done with a soil probe, a thin hollow rod used to explore the soil profile of a site. Geoarchaeologists can use the core sample of a soil probe to identify potential buried layers by past human activity, features below the current stratum, or the extent of an archaeological site. This can also be performed underwater and in wetland environments using a vibracore, which uses a vibration to drive a larger core tube deeper into softer unconsolidated sediments. Friction is reduced through the vibrations, resulting in relatively undisturbed sediment cores compared to other coring methods. By examining natural exposures and investigating the extent of stratigraphy at a site, geoarchaeologists can decipher the evolution of landforms and their effects on the behavior of prehistoric humans and their ancestors. In turn, these insights can underscore the importance of conservational techniques due to the geological processes that affect cultural material left behind.

Applications of geoarchaeology to prehistoric studies has fundamentally reshaped paleoanthropological understandings as the reader has explored throughout the previous sections on hominin lineage in this book. Methods of geochronological dating have resolved many tumultuous conflicts and debates over the past several decades as geoarchaeology gradually found its footing as a distinct discipline. With the more recent innovations in archaeometry and geophysics, archaeologists have become increasingly dependent on interdisciplinary techniques, resulting in the rapid development of geoarchaeological methods in the 21st century. In turn, the theoretical approaches to paleolandscape reconstruction have imparted necessities for revisions to former Eurocentric models of prehistory by dismantling false narratives regarding prehistoric peoples. More importantly, the development of more technologically accessible methods of exploring submerged prehistoric sites, including SCUBA diving, has profoundly contributed to our understanding of submerged prehistoric landsurfaces along continental shelves. Thus, our knowledge of past human migration patterns during glacial and post-glacial conditions has become deeper with the interweaving of disciplines in submerged prehistoric geoarchaeology.

In all fieldwork, identifying the chemical and biological processes of different environments and their stratigraphy can aid investigations into the physical evidence left by past humans. Geoarchaeology offers a specialized approach to fieldwork and laboratory methods by combining multiple disciplines to recognize these processes. Examining the interactions with past humans and their environment at a specific point in time allows geoarchaeologists to reconstruct synchronic landscapes, while studying the evolution of environments over long periods of time and their effect on human settlement patterns aids in reconstructing diachronic landscape models. While the following sections will touch on key concepts regarding sediment and soil-forming processes as they pertain to ancient landscape formation and reconstruction, the reader is encouraged to review these concepts in Michael Waters' *Principles of Geoarchaeology: A North American Perspective*, where this topic is explored in detail. This section will expand on the methodologies and concepts from throughout this book by discussing rock classifications, lithostratigraphy versus pedostratigraphy, soil horizon characteristics and nomenclature, archaeometry and geochemical analysis techniques, geophysics, geoarchaeological predictive modeling, and finally, briefly cover submerged geoarchaeology.

Rock Classifications

Sedimentary

This type of rock undergoes formation processes on or near Earth's outer surface. They are the result of lithified sediment deposits formed when pre-existing rock pieces, minerals, or organic matter become granulated and accumulate. These deposits then undergo compaction and cementation. This can be due to erosion or eluvial deposition of materials that become weathered in place over time. These materials can then be moved by wind and water downslope, or become accumulated due to gravity and eventually lithify into sedimentary rock. This process of physical and chemical changes, known as diagenesis, can take anywhere from thousands to millions of years. The sedimentary rock classification can be further divided into four broad categories based on the processes that led to their distinct formations: clastic, organic, chemical, and biochemical sedimentary rock.

Clastic, or detrital sedimentary rock, is formed when erosional fragments of minerals and other rock deposits (clasts) become lithified. Clastic sedimentary rocks are often grouped based on the particle size of the sediment they consist of. Conglomerate and breccia are coarse-grained rocks formed from clasts of different stones cemented together. Conglomerate rock often exhibits rounded matrices of stone, while in breccia, the geological composition is angular. Sandstone is primarily composed of sand-sized particles, with the most abundant mineral being quartz. Siltstones are made up of smaller particles and are medium-grained sedimentary rocks. They are distinguished from fine-grained sedimentary rocks in their lack of fissility, as they do not have a tendency to split into thin layers. Shale is generally made of clay-sized particles and hardened mud compacted together. These are fine-grained sedimentary rocks that have the smallest particle size of the sedimentary rock classifications.

Organic sedimentary rocks are typically defined as having an increased amount of organic carbon due to the lithification of plant and animal remains. Organic materials in sedimentary rocks are primarily derived from humus, which is the decomposition of plant and animal matter in soil. In oxygen-deficient wetlands like bogs, swamps, and mires, decomposed organic matter known as peat is found in waterlogged conditions. Peat can facilitate the formation of sedimentary rocks like coal, bituminous, and lignite when buried and subjected to compaction and coalification.

In anaerobic salt and freshwater marine environments, organic-rich sediments like sapropel can accumulate from the deposits of decaying phytoplankton, zooplankton, and plant matter. Sapropel is a fine-grained sediment that is often found in anoxic or euxinic conditions where a lower oxygen level at the seafloor inhibits the preservation of organic matter along with an increased concentration of hydrogen sulfide. It contributes to the formation of finely laminated sedimentary rock in benthic faunal areas of marine environments. Along with a reduced diversity of species, high concentrations of trace elements known as geochemical or paleoredox proxies can be found in the laminated layers of these sedimentary rocks as redoximorphic features. These trace elements include uranium, molybdenum, rhenium, and vanadium. These redox proxies can indicate the changes in the oxidation-reduction states of their environment over time, reflected in the microstrata content within these rocks. Geoarchaeologists can use this information to infer the past oxygen conditions of submerged archaeological sites relative to past human settlement patterns.

Chemical and biochemical sedimentary rocks differ in the way geological processes affect their formations. Chemical sedimentary rocks such as calcite, rock gypsum, chert, and travertine form when a precipitate deposits minerals over a long period of time. Biochemical sedimentary rocks such as limestone and chalk are formed from the lithified shells and skeletons of marine organisms.

Metamorphic

This type of rock is formed both deep within the Earth and near the surface when chemical or textural changes alter the mineral composition of igneous, sedimentary, or other metamorphic rocks. This can be due to intense heat and pressure from tectonic events or alterations from hot mineral-rich fluids. This process can be rapid or gradual, taking up to millions of years for metamorphism to occur. Metamorphic rocks are classified as either foliated or non-foliated based on the type of pressure and mineral alignment undergone during formation.

Foliated rocks tend to exhibit a band-like appearance due to the differential stress from direct pressure. When mineral compositions realign under this unequal pressure, they often flatten perpendicularly towards the direction of stress, resulting in layered bands. Types of foliated rock include slate, schist, phyllite, and gneiss. Non-foliated rocks do not exhibit the same banding as foliated rocks due to their formation under relatively uniform low-pressure environments near the Earth's surface. Non-foliated rock can also form when parent rock consists of minerals that do not align easily, such as quartz or calcite, and due to processes like contact metamorphism, where heat from a magma intrusion alters the surrounding geological composition. Examples of non-foliated metamorphic rocks include quartzite, hornfels, marble, and jasper.

Igneous

This type of rock is formed when magma or lava is cooled and solidified. Igneous rocks are classified as either intrusive or extrusive and can be further grouped based on their geological composition and texture. These subgroups are defined as felsic, mafic, and ultramafic igneous rocks. Felsic igneous rocks are typically lighter colored and consist of elements like silicon, aluminum, potassium, and oxygen. Mafic igneous rocks are composed of denser elements such as iron, magnetite, and magnesium, exhibiting a darker color.

Intrusive igneous rocks, also called plutonic rocks, form beneath the Earth's surface as magma slowly cools and solidifies over thousands to millions of years. This creates large coarse-grained, or phaneritic, crystals exhibited in rocks such as diorite, granite, gabbro, and peridotite. Intrusive igneous rocks are generally classified by their relative mineral content of quartz, plagioclase, alkali feldspar, and feldspathoid. Extrusive igneous rocks form when magma erupts at Earth's surface as lava, where it cools and solidifies rapidly, resulting in fine-grained or uniform glassy textures. These aphanitic volcanic rocks contain relatively small crystals compared to intrusive igneous rocks. Igneous rocks that form from these processes include basalt, obsidian, andesite, rhyolite, pumice, scoria, and dacite.

Stratigraphy: Soils & Sediments

In archaeological investigations, stratigraphy is generally prioritized over soil horizons as the method of inquiry when characterizing material deposition chronologies. This is because strata directly correlate to the sequence of chronological events at an archaeological site, allowing archaeologists to establish the evolution of a site and date artifacts accordingly. This is especially true when layers of strata contain cultural remains, which can help geoarchaeologists determine past human settlement activities in relation to geological processes at a site. However, although stratigraphic layers are intrinsically linked to cultural activity, they can become temporally unrelated through periods of disturbances and geological instability.

Strata are defined as the layers of rock, soil, and other materials deposited over time, while soil horizons are the distinctive layers in a soil profile formed by chemical, biological, and physical processes. Soil horizons develop through pedogenic (soil formation) processes involving the chemical and physical weathering of underlying parent material. Parent material is often layers of sediment that have undergone compaction and cementation into hardened layers of consolidated rock over long periods of time. During periods of deposition, sediment transported by fluvial, glacial, eolian, volcanic, and alluvial processes can bury older layers of soil formed from the pedogenic processes of previous underlying sediment deposits. These buried soil layers bracketed by sediment layers are known as paleosols, which represent ancient land surfaces. In contrast, ancient soils that are exposed at the surface are known as relict soils. Sediment deposited by flowing water following human disturbances, like agricultural use and long-term settlement occupation, is known as post-settlement alluviation (PSA), which consists of alluvium and colluvium. Colluvium is poorly-sorted, unconsolidated sediment transported by gravity on slopes.

It is important to recognize and distinguish between the lithostratigraphy of sediment layers representing periods of deposition and the pedostratigraphy of soil horizons representing periods of stability (non-deposition or erosion) in archaeological contexts, as they are likely indicators of when past human activity could have occurred. The boundaries between layers of strata can be identified as conformable, meaning there was a continuous deposition of strata, or unconformable, meaning there was a period of non-deposition or erosion along with potential pedogenic soil-forming processes. Boundaries between layers of strata themselves often represent longer periods of time than the actual matrix deposited. Different sedimentary layers that were not deposited at the same time across their lateral extent are referred to as time-transgressive.

Periods of occupation by multiple different groups represented in archaeological assemblages can become buried in a single depositional event after long periods of stability. This can appear as one coherent matrix in a single stratum, where artifact assemblages can overlie each other in the same paleosol or sediment layers, making it difficult to distinguish between the multiple periods of occupation that took place. Layers of sediment are represented by a prefix numerological and alphabetical system that indicate corresponding strata in a chronostratigraphic sequence. Sediments are classified on the Wentworth Scale by particle or grain size, which ranges in categories from boulders, cobbles, pebbles, sand, silt, and clay.

Soil horizons are defined by their distinct physical, chemical, and biological properties that are usually discerned based on color, texture, and particle size. The presence of silt, sand, loam, or clay particles determines the soil horizon structure and texture. While soil horizons are not always directly linked to temporal events at an archaeological site, they provide geoarchaeologists with an important insight into reconstructing past environments and pedogenic processes. This information can guide the types of methods or techniques used to approach an archaeological site.

Soil color is a key indicator in determining soil drainage and nutrient density, as different geological processes affect the environment. The three components of a soil color are hue, value, and chroma. Hue is the primary spectral color of soil horizons and can indicate soil drainage, mineral matrices, and the presence of organic matter. Value discerns a soil's lightness or darkness, and can be attributed to the content of organic matter as well as its density within the soil. Chroma refers to the saturation or intensity of color in different soil horizons. A lower chroma value, usually appearing muted gray or gray-blue, can indicate the presence of reduced iron compounds common in waterlogged or anaerobic wetland environments. Higher chroma values indicate the potential accumulation or illuviation of iron oxide minerals. More specifically, minerals like hematite contribute to red soils, while minerals such as goethite often yield yellow hues in soil horizons.

The USDA Soil Taxonomy identifies 12 primary orders of soil: Alfisols, Andisols, Aridisols, Entisols, Gelisols, Histosols, Inceptisols, Mollisols, Oxisols, Spodosols, Ultisols, and Vertisols. According to the World Reference Base for Soil Resources (WRB), however, there are 30 major soil groups identified by their formation processes. These groups are Acrisols, Albeluvisols, Alisols, Andosols, Anthrosols, Arenosols, Calcisols, Cambisols, Chernozems, Cryosols, Durisols, Ferralsols, Fluvisols, Gleysols, Gypsisols, Histosols, Kastanozems, Leptosols, Lixisols, Luvisols, Nitisols, Phaeozems, Planosols, Plinthosols, Podzols, Regosols, Solonchaks, Solonetz, Umbrisols, and Vertisols.

O Horizon

Composed of organic matter such as decaying plant material, humus, and peat, this is typically the uppermost soil horizon. This layer is crucial for soil fertility and the redistribution of nutrients from decaying organic matter. It is generally either a few centimeters deep, and in some cases non-existent; however, this can vary depending on the region, with thicker O horizons developing in temperate and tropical forests.

SUBORDINATE DISTINCTIONS

Oi (Fibric) - Slightly decomposed

Oe (Hemic) - Moderately decomposed

Oa (Sapric) - Highly decomposed

Organic Layer

This stratum refers to the O horizon, and can range from a typical thin layer of strata to thicker deposits or even non-existent in some environments. The presence of charred material or carbonization of wood in this layer can indicate past or more recent human activity. It is typically indicative of fresh undisturbed soil; however it can provide clues to landscape stability when buried or disturbed. Thick organic layers represent longer periods of landscape stability without erosion.

A Horizon

This layer is where most biological processes occur and generally consists of a nutrient and mineral-rich soil matrix. It is the layer where most roots grow, and microorganism activity is high. Bioturbation, including faunalturbation and floralturbation, can result from insect, small-animal, and root structure disturbances in the A horizon, which can bring sediments and soils from other horizons into this layer. Burrows created by roots and animals that become filled with materials or soils from surrounding horizons or strata are known as krotovinas, often appearing as distinct features in a matrix. Other types of disturbances frequently occurring in this horizon include cryoturbation (cycles of freezing and thawing) and graviturbation (downslope movement of artifacts and soil by gravity). Sediments composed of windblown materials in this horizon, such as sand or loess, are known as eolian soils.

SUBORDINATE DISTINCTIONS

Apz - Plow zone or agricultural disturbance

PSA - Post-settlement alluviation

Aa - Highly decomposed organic matter

Ae - Organic material of intermediate decomposition

Ai - Slightly decomposed organic material

Ag - Gleying, indicating a waterlogged condition

Ak - Accumulation of carbonates

An - Accumulation of sodium

Ar - Partially weathered or soft bedrock

Ah - Accumulation of humus (mollic)

Topsoil

This stratum corresponds to the A horizon and can help archaeologists understand landscape or environmental changes over time based on new depositional materials such as alluvium. The presence of a buried A horizon, sometimes referred to as dark earth, signifies that a new layer of sediment or material has accumulated over the former paleosol during a period of deposition. This sediment is often deposited by flowing water following human disturbances, like agricultural use and long-term settlement occupation, which is known as post-settlement alluviation (PSA). PSA consists of alluvium and colluvium. Colluvium is unconsolidated sediment transported by gravity, which accumulates downslope in a poorly sorted matrix. When referring to an A horizon that has been disturbed by agricultural means, this stratum is often called the Ap horizon or plow zone (Apz).

E Horizon

The E horizon is typically composed of coarse-grained sediments with less clay content and an overall lighter color than the A horizon. This is a result of leaching, or eluviation, due to the downward movement of water washing out clay, iron content, and organic matter from this layer. This process leaves behind heavier sand, silt, and quartz particles. The leached material forms illuvial deposits that accumulate in lower subsoil horizons. These horizons are commonly present in coniferous forests, often appearing as lightly colored coarse leached layers of soil.

SUBORDINATE DISTINCTIONS

Ea - Accumulation of aluminum

Eg - Gleyed eluvial horizon; reduction and removal of iron due to saturation with stagnant water

Eh - Accumulation of humus

Ei - Slightly decomposed organic matter

Ex - Dense brittle layer within the eluvial horizon (fragipan)

Eb - Buried eluvial horizon

Eluvial Layer

This stratum coincides with the E horizon and is frequently present in forested regions where there is heavy rainfall and parent material containing high concentrations of sand. As water percolates downward through a soil profile, it carries dissolved minerals, ions, small particles, and organic matter which is removed from upper layers. These particulates accumulate in the subsoil as illuviated materials. Geoarchaeologists can infer how archaeological sites are formed, occupied, and subsequently buried by examining eluvial strata. These layers provide key insights to the history of soil forming processes at a site by indicating whether the soil was stable over long periods of time or subject to a disturbance. For example, stable landscapes and forest soils with developed root structures are more likely to exhibit distinct eluviation. Eluviated layers may also indicate former agricultural activities based on the depth, thickness, and consistency of the material left behind from the leaching of water. When a heavily-leached, mineralized eluvial layer is present, it may be the result of poor nutrient recycling indicating the abandonment of a site.

B Horizon

An accumulation of leached materials from the A and E horizons contributes to the formation of this horizon, known as subsoil. Silicate clay, aluminum, iron content, and other trace minerals are deposited in a process known as illuviation, which contributes to the composition of the B horizon. While primarily consisting of high clay content, higher concentrations of organic matter deposited through illuviation can increase cation exchange capacity (CEC). This refers to a soil's capacity to hold plant nutrients in the form of positively charged ions like calcium, magnesium, and potassium, known as cations, which are attracted to and held together by negatively charged clay particles. When anaerobic (low-oxygenated) conditions develop in the upper soil profile due to saturation, the reduction of iron and manganese can lead to a blueish-gray B horizon. This may be followed by the presence of a water table known as hydric soil.

SUBORDINATE DISTINCTIONS

Bh - Accumulation of illuvial humus (organic matter)

Bt - Accumulation of clay (argillic horizon)

Btv I, II, III (Plinthic) - Accumulation of silicate clay and plinthites; an iron-rich, humus-poor mixture of clay with other minerals that forms concretions after repeated drying

BE - Transitional layer between the E and B horizons

BA - Transitional layer between the A and B horizons

Bw - Weakly developed subsoil horizon with altered color or structure

Bg - Gleyed subsoil with poor drainage

Bn (Solonetzic) - Subsoil with columnar structure and high sodium content

Btn - Subsoil with both characteristics of Bn and Bt (argillic and high sodium content)

Bhs - Accumulation of organic matter along with iron and aluminum oxides

Bo - Accumulation of iron and aluminum oxides

By - Accumulation of gypsum (gypsic)

Bk - Accumulation of calcium carbonate (calcic)

Bo - Residual iron and aluminum concretions (oxic)

Bq - Accumulation of silica

Bw - Weakly developed subsoil horizon with altered color or structure and a loss of carbonates with no illuviated accumulation of concretions (cambic)

Bz - Accumulation of salts that are more soluble than gypsum

Subsoil

Considered sterile strata in archaeological investigations, this layer is attributed to the B horizon subsoils and is not typically expected to contain many remnants of past human activity. While sometimes containing deeply buried features or artifacts, this stratum primarily contains clay or rock particulates formed over long periods of time without much human interference. Subsoils may also still contain bioturbation and deep root impasses, but they are typically less disturbed by wind or water processes. When water does enter clay-rich subsoil, it can cause swelling with subsequent drying and cracking known as argilliturbation. This can cause soil and materials from surrounding horizons to enter these cracks, disturbing the subsoil matrix. Subsoil strata often exhibit red or yellow hues from illuviated clay and iron oxide deposits.

C Horizon

The C horizon serves as a transitional layer between bedrock and the upper soil horizons. Weathering processes occur less intensely at this horizon, and it is primarily composed of loosely packed minerals with organic matter known as unconsolidated parent material. There are typically small amounts of organic material and minimal root development in the C horizon.

SUBORDINATE DISTINCTIONS

Cr - Weathered or soft bedrock

Cd - Dense horizon with restricted root impasses

Cv (Plinthic) - Plinthic concretions

Cg - Prominent mottling due to fluctuating water table

Ck - Accumulation of carbonate or calcium carbonate

Cm - Cementation occurring

BC - Transitional layer between B and C horizons

CB - Transitional layer between C and B horizons

AC - Transitional layer between A and C horizons

CA - Transitional layers between C and A horizons

Substratum/Parent Rock

This stratum corresponds to the C horizon and can contain a number of clues about an archaeological site's location and its preservation. Fragments of bedrock and other geological deposits found in parent rock strata can provide details about the environment in which a site was formed. Geoarchaeologists can use this information to determine how the surrounding geological processes influenced settlement patterns. While the substratum is not a cultural layer, by studying its properties and depth, archaeologists can understand the soil formation processes occurring at a site. It may also be indicative of the environment and geological processes that influenced human settlement and activity in a particular area.

R Horizon

Made of bedrock or consolidated rock, these horizons are at the bottom of the soil profile. They are usually composed of unweathered or partially weathered hard rock parent material. These horizons are not soil material but rather the underlying geological formation that has been weathered. There are no subordinate distinctions for this horizon.

Bedrock

This stratum is composed of unweathered or partially weathered parent material made from sedimentary, metamorphic, or igneous rock. It underlies the surface soil horizons and broken rock known as regolith. Bedrock can be exposed at the surface of geological formations or buried deeply underground. Exposed bedrock features, including grooves, mortars, and grinding surfaces, can provide insight into the food processing or resource use of past human settlements.

Additional Subordinate Distinctions

a - Highly decomposed organic material

b - Buried horizon

c - Concretions present

e - Intermediate decomposition of organic material

f - Frozen soil with permanent ice

g - Gleyed horizon

h - Illuvial accumulation of organic matter

i - Slightly decomposed organic material

k - Accumulation of calcium carbonate (calcareous)

m - Indurated (hardened) soil horizon with 90% or more cementation occurring (further designated by km for accumulation of carbonates and ym for accumulation of gypsum)

n - Accumulation of sodium

p - Plowing or other disturbance

s - Illuvial accumulation of sesquioxides

t - Accumulation of silicate clay

v - Accumulation of salts more soluble than gypsum (salic)

w - Weakly developed layer with some evidence of alteration

x - Dense brittle layer (fragipan)

Stratigraphy: Soils and Sediments References:

Waters, Michael R. 1992. *Principles of Geoarchaeology: A North American Perspective*. University of Arizona Press: 45-53.

Soil Science Society of America. 2022. "Designations for Soil Horizons and Layers." In *Glossary of Soil Science Terms*, Appendix II: 146-149. SSSA.

Archaeometry & Geochemical Analysis

As modern technology continues to innovate, geoarchaeology rapidly evolves from different branches of scientific inquiry. Geoarchaeologists approach landscapes with the insight of an interdisciplinary toolkit to reconstruct the long-term effects of environmental changes on past human settlements through techniques from archaeometry and geochemistry, alongside utilizing geophysical technology. To better understand the depositional strata of an archaeological site, geoarchaeologists not only incorporate these techniques in the field but also as a fundamental aspect of laboratory investigations. Geochemical analysis allows for the precise determination of the compositional makeup of geological and cultural materials.

While basic knowledge of chemistry is a crucial aspect in determining the biological processes affecting an archaeological site, along with the chemical consolidation of its artifacts, physics is a cornerstone of geoarchaeological methods in the laboratory. This is especially true when considering light as a stream of photons in applications such as spectrophotometry, photoluminescence spectroscopy, portable X-ray fluorescence (pXRF), Fourier transform infrared spectroscopy (FTIR), and scanning electron microscopy (SEM), among other approaches. Charged particle interactions with cultural material or geological formations can derive accurate interpretations of the geochemical compositions of an object.

Redox-sensitive trace metals (RSTMs) are elements known in archaeological contexts as geochemical or paleoredox proxies. These trace elements include uranium, molybdenum, rhenium, and vanadium. Paleoredox proxies can indicate the changes in the oxidation-reduction states of marine or terrestrial wetland environments over long periods of time, which are analyzed in microscopic sediment deposits. Geoarchaeologists can derive this information from tools like spectrophotometry, which analyzes the absorption factor of light particles as they interact with different materials. This can provide data about the past oxygen conditions of a submerged archaeological site and the geochemical composition of different remains or sediment profiles. The density and spatial distribution of RSTMs can determine the concentration of nutrient cycling in different sediments over long periods of time. The oxidation of minerals such as iron and manganese alters the binding processes of organic carbon, which can impact the availability of nutrients and organic particles.

The subdiscipline of archaeometry and the techniques derived from geochemistry play important roles in modern geoarchaeological investigations. The following brief sections will introduce some common and fairly uncommon methods derived from both archaeometry and geochemistry to the reader as a general introduction. These topics should be further explored and properly learned based on the necessity of their application, and these sections should not be used as a manual to operate or interpret with these methods. It takes many years of rigorous study and practical application to not only properly use some of the following instruments, but to operate a number of them under the specific safety criteria required. Interpreting any archaeological material using these methods also requires an experienced user who understands the context in which they are analyzing. Thus, the discipline of geoarchaeology offers a culmination of specialized approaches to investigate these contexts.

Particle Size Distribution

Also known as granulometry, it is a technique used in sedimentology to measure the range and proportions of different particle sizes within a soil horizon or aqueous solution. This property can be measured through sieving, laser diffraction, or using a hydrometer. It provides geoarchaeologists with insight into the depositional history of a site, which can help determine potential occupation periods based on the ranges of paleoredox proxies or carbonized deposits present.

Total Dissolved Solids

As opposed to Particle Size Distribution, Total Dissolved Solids (TDS) testing encompasses the entirety of organic and inorganic materials present in a solution. This is usually measured in parts per million (ppm), and can be taken using an electronic TDS reader. While the total concentration of dissolved solids can be accurately measured in fresh or mildly brackish water, salt water contains particle counts topping over 32,000 ppm. This can cause inaccurate measurements or errors in electronic TDS readers, as they are typically calibrated up to 9999 ppm. For salt water or aqueous solutions with higher salinity, a conductivity meter can measure the concentration of salt present based on the flow of electric charge. This is because solutions with a higher salt content have more conductivity due to the increased amount of charged ions present.

Specific Gravity

The total dissolved solids within a solution can also be tested using a refractometer, which can take the specific gravity of a solution and be calibrated for either parts per million or parts per thousand measurements. This is done by sampling the refractive index of a solution, which is associated with the concentration of total dissolved solids. The greater the refractive index, the higher the concentration of dissolved solids. This is especially useful for saltwater applications when the upper calibration limit of electronic TDS testers cannot produce accurate readings and are susceptible to corrosion.

Specific gravity is calculated using the relative density formula, which is the density of the solution divided by the density of water at 4°C (32°F):

$$RD = \frac{\rho substance}{\rho reference}$$

Density of water at 4°C ($\rho water$) = $1000\frac{kg}{m^3}$

$$SG = \frac{\rho solution}{\rho water}$$

$$SG = \frac{\rho solution}{1000\frac{kg}{m^3}}$$

Micromorphology

To determine the occupation conditions of coastal or wetland archaeological sites, geoarchaeologists and micropaleontologists can identify foraminifera morphologies characteristic of intertidal zones within a sediment. This provides insight into the relationship between sea level and the occupation of these sites. The different geological ranges of foraminifera species make them useful for dating the relative age of archaeological strata or sedimentary depositions.

Foraminifera are single-celled organisms that are members of a protist phylum that commonly exhibit an armored shell. Their earliest forms date back to the Cambrian Period, and the presence of specific species can indicate whether a site was once a marsh, wetland, or completely submerged environment. Soil probing and taking thin sections of undisturbed sediment are common techniques used to sample foraminifera under a compound light or SEM microscope.

Flotation Sampling

Flotation samples collected from archaeological deposits generally consist of cultural debris (ash, charcoal, flora, fauna, burned rock, and small artifacts) mixed with or without sediment and other natural organic and inorganic inclusions. The samples are often collected from thermal features such as hearths and roasting pits or other features that have a high potential to contain cultural materials associated with the occupation of the site. Specimens recovered from flotation samples can include the remains of plants, animals, and insects, in addition to small stone tool debitage and pottery sherds.

Palynology

Pollen is a coarse powdery substance created by plants during their reproduction processes. It can be transported by wind and through human activity. Phytoliths are microscopic silica structures produced by plants in and around cellular walls and are deposited into soils and sediments as plant remains decompose. They are distinctive in size and shape and are decay-resistant. Phytoliths typically become incorporated into sediments at the location of plant death and decay; however, they can be transported through human activity, animal feces, and soil deposition. Analysis of pollen and phytolith remains can provide valuable insight into paleoenvironments and past human subsistence or plant processing techniques. Samples can be collected from pits, living area surfaces, middens, agricultural fields, ground-stone surfaces, ceramic vessels, and stratigraphic columns. Pollen and phytoliths are commonly collected in the field in a combined sample.

Once collected, unwanted carbonates and silicates are dissolved from the sample, typically using hydrochloric or hydrofluoric acid, while strong bases such as potassium hydroxide break down excess organic matter. The clean sample is mounted on a slide and stained to improve visibility when analyzed under a microscope. Either a high-powered compound light microscope or a scanning electron microscope (SEM) is used to examine the morphology and number of pollen particulates (palynomorphs) present to build a statistically representative model of the sample. By comparing pollen assemblages from different layers of a sediment core, the vegetation and climate of a paleolandscape can be reconstructed to determine how it may have influenced past human settlement and activity in an area.

Fourier Transform Infrared Spectroscopy

To analyze the chemical composition of different artifacts, geological formations, and even sediment profiles, geoarchaeologists can use Fourier transform infrared spectroscopy (FTIR) to expose samples to different wavelengths of infrared light. This provides geochemical data about the material's infrared light absorption patterns that correspond with the vibrations of its chemical bonding and molecular structure. The data is displayed as an infrared spectrum plot, revealing the fingerprint of functional groups and molecular bonds present in the sample.

Mass Spectrometry

To determine the molecular weight or identify compounds in a substance, mass spectrometry vaporizes a sample and exposes those particles to an electron beam. This removes electrons and creates positively charged ions. The resulting ions are accelerated in an electric field and subsequently separated in a magnetic field based on their mass-to-charge ratio (m/z). A detector measures the separated ions, which provides a spectrum based on m/z values that correspond with specific ions present. This data can be analyzed to identify and determine the molecular weight of compounds present in inorganic molecules and sediments. The force and deflection experienced by ions passing through the electric and magnetic fields of a mass spectrometer are dictated by the Lorentz Force equation:

$$F = q(E + v \, x \, B)$$

Where:

q = a particles charge

E = the strength of the electric field

v = the velocity of a charge particle

B = the strength of the magnetic field

Different applications of mass spectrometry exist to analyze a vast array of molecules, including DNA, RNA, metabolites, and lipids. Zooarchaeology by mass spectrometry (ZooMS) focuses on Type I collagen peptides, which are proteins found in animal tissues like bone, ivory, antler, and hide.

This process of collagen fingerprinting is used to identify different taxa of animals based on their remains within archaeological or paleontological contexts.

Accelerator mass spectrometry (AMS) uses a tandem accelerator to boost ions to much higher energy levels than normal mass spectrometry is capable of handling. A tandem accelerator can achieve higher energies than single-stage accelerators due to a two-stage acceleration process. First, a negative ion source is given a charge to produce negatively charged ions, which gain energy as they accelerate towards a high-voltage terminal. Upon reaching the terminal, the ions pass through a stripping barrier, typically a thin carbon foil or, in some setups, a gas, which removes electrons making them positively charged. These positively charged ions become repelled by the high-voltage terminal, thus gaining velocity and additional energy through a second acceleration. The accelerated positive ion beam is then directed towards a sample. This aids in the separation and detection of rare isotopes from those more commonly found. Because AMS offers higher sensitivity and selectivity for the detection of rare isotopes like carbon-14 and beryllium-10 at ultratrace levels, it is ideal for analyzing small samples, making it a choice method for radiocarbon dating.

Spectrophotometry

Offering another non-destructive archaeological technique, spectrophotometry can also identify the composition of artifacts or geological material. It measures the absorption or transmission of the visible and UV light spectrums to determine the concentration of different compounds. Unlike FTIR, spectrophotometry usually requires dilute solutions of a sample to be accurately measured. The data is also displayed as a wavelength spectrum plot, revealing the fingerprint of functional groups and molecular bonds present in the sample across varying spectra.

Photoluminescence Spectroscopy

This method involves using different wavelengths of light to excite organic compounds to higher energy levels. This reveals phosphor signatures that indicate the presence of pigments, carbonization, and mineral inclusions, among other trace element staining. For optimal excitation wavelength ranges, it is generally ideal to match the material's absorption bands. Fluorescence excitation typically falls within the UVB and UVC ranges, between 250 and 350 nanometers, while visible light up to 700 nm is commonly used based on the material being measured.

X-Ray Absorption Spectroscopy

Unlike photoluminescence spectroscopy, X-ray absorption spectroscopy (XAS) uses X-rays to excite electrons and measure their absorption by different materials to identify elemental composition and chemical properties. XAS is primarily used to identify the type of pigments used in prehistoric pottery or artwork, and estimate the firing temperature of pottery based on changes in mineral phases within its structure. The data is presented as an X-ray spectrum graph that plots the absorption of X-ray radiation by the sample at different energy levels.

Portable X-Ray Fluorescence

Used to identify the presence and quantity of different elements in archaeological or geological material, portable X-ray fluorescence (pXRF) is often used in the field for a quick and non-invasive analysis. The device emits low-energy X-ray beams to knock electrons out of the inner shell of a sample's atoms. This forces an electron from a higher-energy shell to take its place, which emits a fluorescent X-ray photon in the process. The device measures the energy levels of these emitted photons, which are represented by a spectrum plot correlating the varying energy levels by elements present in the sample.

X-Ray Diffraction

To investigate the structure or properties of crystalline materials, X-ray diffractometers (XRD) are a useful tool to determine atomic arrangements. These devices use high-energy electron beams to generate X-rays, which are then scattered by different arrangements within a crystal lattice. When these scattered X-rays interfere with each other constructively, they produce a strong signal and are considered in phase. When the scattered X-rays interfere destructively, they are out of phase, producing a weaker signal. This diffraction pattern can reveal the structure of crystalline samples, essential for the study of geological materials. Bragg's Law defines the conditions under which X-rays constructively interfere, accounting for destructive interference as these X-rays will not produce detectable diffraction patterns:

$$n\lambda = 2d\ sin\theta$$

Where:

n = integer representing order of reflection

λ = wavelength of incident X-rays

d = distance of reflecting planes in the crystal lattice

θ = X-ray angle of incidence relative to the reflecting planes.

Scanning Electron Microscopy

As opposed to compound light microscopy or stereoscopy, scanning electron microscopy (SEM) uses a focused electron beam to produce high-resolution images of samples, usually down to a few nanometers. A narrow electron beam makes contact with the surface of a sample, causing the sample itself to emit secondary or backscattered electrons. A detector then measures the number of ejected electrons, where the number counted determines the brightness of each pixel appearing on a monitor. This process continues until the beam has covered the entire area to be imaged. While many optical microscopes are capped at 500x to 1500x magnification, some SEMs can reach up to 1,000,000x magnification. This not only reveals the composition and structure of materials, but their topography as well. This data is crucial for identifying organic pigment traces in pottery, the structure of osseous material, and revealing wear patterns on stone tools.

Radiocarbon Dating

Also referred to as carbon-14 dating, this method has been widely adopted since its inception in the 1940s to determine the ages of organic material or remains. The technique involves measuring the decay of residual radioactive carbon-14 present in a sample, which has a half-life of approximately 5,730 years. This is done typically through an accelerator mass spectrometer

(AMS), which uses a particle accelerator to separate ions as opposed to a magnetic field.

As cosmic background radiation enters Earth's atmosphere, some of it interacts with nitrogen-14 atoms. This interaction produces carbon-14, which reacts with ambient oxygen to form carbon dioxide, known as the radiocarbon cycle. Throughout an organism's life cycle, it constantly builds carbon-14 levels in its tissues from carbon dioxide. At death, the absorption ceases, and carbon-14 begins to decay. The remaining amount of carbon-14 indicates how long ago the specimen died. This is true for any material from formerly living organisms and applies to organic materials like wood, charcoal, seeds, bones, shells, hair, fabrics, and animal hides. Radiocarbon dating can accurately date this material up to around 60,000 years ago, as residual carbon-14 is too low to measure against cosmic background radiation beyond this limit.

While radiocarbon dating has been a reliable method for terrestrial contexts, marine samples often return with older dates than the actual age represented by the specimen due to the marine radiocarbon reservoir effect (MRE) (Alves et al. 2018). This offset is due to the fact that carbon in deep ocean water is older than carbon in the atmosphere. Varying factors from ocean paleoecology, current systems, and eustatic sea level changes due to glacial and inter-glacial periods have made it difficult to quantify the effect of the MRE without local calibrations. By quantifying local deviations from the global average

MRE, the effects on a sample can be accounted for. It can also be used to determine past and present changes in ocean circulation patterns relative to carbon exchange.

Beryllium-Based Radiometric Dating

Unlike radiocarbon dating, this technique is primarily used for the dating of geological materials that absorb beryllium-10 through cosmic rays. Beryllium-10 is a radioactive cosmogenic nuclide produced in the atmosphere and in minerals like quartz by the interaction of cosmic rays with oxygen and silicon atoms. The half-life of beryllium-10 is approximately 1.387 million years, allowing accurate dating of specimens up to around 10 million years old. Beryllium does not form strong bonds with organic matter; thus, it is found primarily in geological formations and wetland sediments. It provides insight into depositional events and the extent of geological periods. This dating method measures beryllium-10 and beryllium-9 isotopes to determine the exposure age of rocks and sedimentary deposits at Earth's surface or the duration of their burial. An exposure age can be determined by measuring the concentration of beryllium-10 isotopes in a sediment sample, which are produced when cosmic rays interact with the oxygen and silicon atoms present in rocks or soil. The stable isotope of beryllium, beryllium-9, which is naturally abundant in Earth's crust, acts as a normalization factor to account for variations in the delivery of beryllium-10 to the sample from the atmosphere or mineral matrix.

Argon & Potassium Radiometric Dating

The two most commonly used argon dating methods in geoarchaeology are potassium-argon (K-Ar) dating and argon-argon dating. Potassium-argon dating measures the ratio of radioactive potassium-40 to its decay product, argon-40, particularly in volcanic rocks where argon gas accumulates as they cool and solidify. By measuring the amount of potassium-40 and argon-40 within a sample against the decay rate, the time since the rock layer last solidified can be determined.

Argon-argon (40Ar/39Ar) dating is a relatively refined version of potassium-argon dating that is based on the decay of radioactive potassium into radioactive argon. It is also suitable for dating volcanic rock or ash layers and uses neutron irradiation to convert potassium-39 to argon-39, allowing for a single measurement of argon isotopes to determine the age. This circumvents probable issues with potassium-argon dating due to it requiring separate measurements of potassium and argon, which can introduce potential uncertainties.

Thermoluminescence Dating

This dating method measures the amount of electrons emitted from a heated sample, which indicates the time since the sample was last heated or exposed to sunlight. Minerals such as quartz and feldspar, which are found in pottery, sediments, and lithics, contain imperfections in their crystal structure known as electron traps. As these minerals become exposed to naturally occurring ionizing radiation from other minerals, such as uranium, potassium, and thorium in the surrounding environment, this radiation ejects electrons from their atoms. Some of these electrons become trapped in the crystal structure, which can then be reheated and measured to determine the age of a sample. It is particularly useful for dating pottery, as electrons become freed from crystalline structures once the pottery is fired. This allows for the new accumulation of electrons from surrounding radiation over time, which allows pottery to be dated accurately to when its material was last heated with thermoluminescence.

Fission-Track Dating

Fission-track dating is a radiometric technique that analyzes the density, length, and number of damage trails left in the wake of uranium-238 fission within crystalline structures. The radioactive isotope uranium-238 has the potential to decay through spontaneous fission, which releases high-energy fission fragments. These fragments create fission tracks as they travel through the crystalline structure; however, these tracks can be erased by heat, making this method sensitive to any thermal history of a sample. By knowing the uranium concentration within the sample, the age since it first cooled can be estimated based on the number of fission tracks present and their properties.

Geophysics

While physics is an important aspect of the laboratory to investigate the composition of sediments and objects, it is equally a fundamental principle of data collection in the field. Geophysical approaches to archaeology, or *archaeogeophysics*, have modernized the way fieldwork is performed by streamlining the process of archaeological inquiry. This includes the use of GIS probability-based predictive modeling, georeferencing aerial or historic maps, remote sensing methods, and magnetometry to better reconstruct paleolandscape environments and interpret the data once fieldwork is completed.

The use of digital technology in the field for geophysical acquisition has become a standard approach to cultural resource management. GIS map models paired with GNSS receivers provide sub-meter tracking for accurate archaeological data collection. By incorporating preemptive geophysical approaches to investigate these potential archaeological sites, archaeologists can take advantage of transect pre-plots, high, medium, and low probability areas, well or poorly-drained soil data, along with a suite of tools to plot new data in the field. This creates professional and detailed interpretations of archaeological contexts that make revisiting sites for further investigations more straightforward.

More crucially, geophysical applications to archaeology have provided non-destructive techniques for investigating subsurface features, geological deposits, and buried archaeological materials. This has transformed approaches to long-term and multiple-phase fieldwork sessions by providing archaeologists with insights into the depositional history of an area they intend to excavate. By incorporating geophysical methods such as ground-penetrating radar or magnetometry, areas of excavation that are more promising for data collection can be identified, thereby maximizing the efficiency of research methods. These applications aid in locating the nature and extent of buried remains, which can help in the protection and management of a site by minimizing disturbance and preserving the site for future research.

Ground Penetrating Radar

To investigate potential subsurface features, artifact deposits, or distinguish between changes in soil horizons based on possible human activity, ground penetrating radar (GPR) provides a viable method of geophysical survey with certain limitations. GPR sends short pulses of electromagnetic signals into the ground and collects data from signals that are reflected back by changes in the electromagnetic properties of the subsurface. These changes can include different materials, structures, or soil layers. A 3D map of these subsurface features can be created based on strength, shape, and the time it takes for electromagnetic signals to return. While GPR provides geoarchaeologists with a way to approach archaeological sites by detecting contrasts in subsurface materials, it cannot differentiate the material composition of these contrasts. Furthermore, clay-rich soils and the presence of groundwater can inhibit the effectiveness of GPR by absorbing radar signals.

LiDAR

While GPR is primarily used for subsurface imaging, LiDAR (light detection and ranging) is a form of remote sensing that focuses on creating 3D maps of an environment's surface. LiDAR emits short near-infrared laser pulses on the surface and measures the time it takes for them to return to the sensor. In hydrological applications, green-light lasers can be used to penetrate water columns and provide a view of bottom elevation levels or depth. This data is used to create high-resolution 3D maps of structures, bathymetry, and geological formations in the field; however, it is unable to penetrate rock and dense soils.

Magnetometry

The use of magnetometry in geophysical applications to archaeology provides a way to detect and map subsurface variations in Earth's magnetic field, which can reveal buried features and magnetic anomalies (El-Qady, Metwaly, and Drahor 2019, 3-4). The most commonly used for archaeological surveys are cesium, proton, and fluxgate magnetometers. These devices can detect subtle differences in the strength and direction of magnetic properties in sediment due to buried objects, structures, pits, and hearths. They can also indicate the presence of redoximorphic features from magnetic minerals that are sensitive to changing reduction-oxidation conditions. This is crucial for paleolandscape reconstruction when investigating potential past settlements and deciphering long-term environmental changes that affected the site.

Magnetic Gradiometry

This geophysical technique for archaeological survey uses multiple magnetometers to measure magnetic gradients. This provides more accurate information about subsurface magnetic anomalies like hearths, foundations, burials, and pits than standard magnetometry. Buried features cause micro-changes to the Earth's magnetic field due to alterations in soil composition and the presence of heat-treated rock. Gradiometry measures these magnetic signatures to interpret the location of submerged historical and paleolandscape features. The data is then presented as a grayscale image indicating field strengths, where null readings are medium gray, positive readings are dark gray to black, and negative readings are pale gray to white.

Predictive Modeling

Predictive modeling refers to the various approaches that attempt to model features on a landscape in an effort to predict the locations of previously unidentified archaeological resources and sites. Probability-based predictive models have a set of assumptions regarding landscape features, which is the data set used to derive high, medium, and low probability areas. When building predictive models, archaeologists generally work under the assumption that past human settlements preferred well-drained soils within a relatively close distance to permanent freshwater sources. Classifying soil data based on drainage, representing areas with access to freshwater within specific distance parameters, and using nearby cultural resource locations from a State Historic Preservation Office or other reliable database when building a predictive model can create more accurate probability polygons.

Many GIS predictive models also include elevation data, such as slope or aspect. Slope is recorded as an angle, and aspect is the cardinal direction that a portion of a landscape is facing. For water feature data, the national hydrography data set from USGS can provide files for flowlines, bathymetry, and waterbodies in the United States. Soil drainage data can be accessed through the Web Soil Survey website by downloading soil data based on the area of interest. This tabular data can then be aggregated in a database and defined based on drainage type.

Incorporating GIS software capable of constructing geospatial models allows geoarchaeologists to forecast the potential spatial distribution of past human activity patterns. This can streamline the flow of archaeological excavations by combining predictive models with preemptive geophysical survey data to better understand and investigate sites before breaking ground. Archaeologists can use GNSS receivers to enable live tracking for accurate plot positioning by simultaneously connecting multiple users into the same collection databases while in the field. This can ensure more precise collection of coordinate-specific survey points and site information when proceeding with multiple phases of fieldwork.

Among the predictive model applications to determine probable archaeological sites, they can also be used to create geoarchaeological predictive models. Stratigraphic predictive models forecast the spatial distribution of strata by simulating their formational processes over time. Geomorphic predictive modeling simulates the behavior and evolution of landscapes at Earth's surface to determine how different features were formed. These models can be built using geomorphological maps, which describe landform features and geological processes, and are also used to create or refine existing ones.

Vector Data

Vector data is used to represent points, lines, and polygons, which are symbolized by geometric shapes within the layers of the GIS environment. These are used to create probability areas and transect points, allowing for precise spatial representation and analysis. Each vector layer is defined by a set of coordinates associated with a Coordinate Reference System (CRS), which tells the GIS model where the data is located on Earth and how it should be displayed.

Raster Data

Raster data is a continuous data set that contains values across a raster layer's extent. It represents data sets like elevation, temperature, or rainfall. The values are considered representative of any given area and determine continuous phenomena across a landscape. Raster data used to orient points, lines, polygons, or data used in a larger spatial context can act as a basemap layer for GIS models.

Coordinate Reference Systems

A Coordinate reference system (CRS) describes the generalized shape of the Earth based on an ellipsoid model. These are used to represent the locations of features within a common mathematical representation of an area. An ellipse is never completely perfect due to the bulge at Earth's equator and flattening of its poles. Many coordinate reference systems exist to create more accuracy at a localized scale. They are related to a datum, which defines the origin and orientation of the coordinate axis. There are two types of coordinate reference systems often encountered when working with GIS software: geographic and projected coordinate systems.

Geographic Coordinate System

This type of coordinate reference system represents the world in three dimensions and provides a more accurate model of the Earth's surface. Common examples include latitude and longitude, often expressed as either degrees, minutes, seconds, or decimal degrees.

Projected Coordinate System

More frequently used in archaeological GIS applications, a projected coordinate system represents the world in two dimensions. They support most models of geospatial analysis that also rely on linear measurements. Ensuring your area of interest is in the correct projected coordinate system before building any predictive model will prevent conflicts of coordinate systems when incorporating soil drainage, hydrography, and cultural resource data.

Submerged Geoarchaeology

As geoarchaeology continues to pioneer advanced technology into cultural resource management from a multitude of scientific disciplines, maritime archaeology reaps the benefit of these geophysical innovations underwater. Applications to submerged subsurface data interpretation, including magnetic susceptibility, acoustic radar, sub-bottom profiling, and seismic data collection, have redefined the industries involved with maritime science and the research of submerged prehistoric landscapes as a whole. Maritime archaeology now incorporates the use of geospatial analysis on the water with the novel and robust technology of geophysics to study underwater cultural remains and sites, along with the paleolandscapes that once shaped their deposition.

Periods of marine transgression and regression influenced the spatial distribution and preservation of coastal prehistoric archaeological sites during the Early Holocene, when postglacial conditions contributed to rapid eustatic sea level rise. Factors such as land uplift due to the removed weight of Pleistocene glaciers, known as isostatic rebound (Lemke 2024, 32), changed coastlines and altered access to subsistence strategies such as fishing due to landscape instability and saltwater intrusions. These effects on prehistoric sites located on now submerged landmasses along continental shelves provide paleoenvironmental data which directly informs archaeological theory by raising questions concerning differentiations between culturally significant from purely environmental when investigating paleolandsurfaces (Bicket et al. 2014, 224-225). In this way, geoarchaeology can aid in describing the significance of geological formations that provide a deeper, often hidden context in individual sites, which can contribute to identifying and reconstructing the submerged paleolandscape that prehistoric peoples may have had a presence on.

While maritime archaeology can be discussed as a much broader topic encompassing interdisciplinary approaches to study submerged sites, this section will be primarily focused on geophysics and its implications for the subfield. It should be noted that the subsections on archaeometry and probability-based predictive modeling are also implicit in maritime archaeological methods, as chemistry, biology, and physics are fundamental aspects of maritime field and laboratory work. The techniques used to investigate redoximorphic features and magnetic anomalies, identify compositions of materials, and chemically consolidate cultural material are adapted similarly from methods of terrestrial geoarchaeological inquiry.

Maritime predictive modeling applications that simulate past human settlement patterns and seafloor geology can aid in determining the locations of submerged prehistoric sites.

They can also help reconstruct sea-level curves on a graph, which demonstrate periods of eustatic sea level rise on Earth. In North America, many prehistoric Paleoindian and Early Archaic sites along the Atlantic and Gulf coastlines are now submerged underwater. This is due to the rise in sea level corresponding with glacial melt of the Last Glacial Maximum (LGM) that submerged the broad and gently sloping Atlantic and Gulf continental shelves. The prehistoric shoreline of the Atlantic and Gulf coasts would have been displaced by 100 to 150km from their present-day configurations, opposed to the steeply inclined narrow shelf of the Pacific coast, which would have only been displaced by roughly 0.5 to 10 km (Waters 1992, 262-269). The use of predictive modeling in maritime contexts combined with geophysical instruments is a powerful approach for locating these submerged prehistoric sites.

Magnetic Susceptibility

Magnetometry has quickly become one of the most applied geophysical techniques to survey for magnetic anomalies underwater. Cesium magnetometers use a vapor cell filled with cesium gas through which a laser is emitted. When a magnetic anomaly is present, the cesium atoms fluctuate, causing the wavelength of light they absorb to change. A detector measures these changes based on the strength and direction of the magnetic anomaly.

Cesium magnetometers are typically deployed during maritime surveys and require the device to be placed in a specialized waterproof cylinder. It is then towed behind a vessel at a relative enough distance to prevent interference from any magnetic fields the ship or devices onboard emit. The depth of tow is determined by the goal of the survey, and whether the presence of ferrous materials requires lower sub-bottom investigations on the seafloor.

Fluxgate gradiometers are also deployed for similar purposes and provide higher-resolution imaging for shallower archaeological surveying. This is due to their higher sensitivity to weaker magnetic fields that are more pronounced in shallower depths. Their tendency to collect background noise and experience drift in more difficult weather or environments makes them less suitable for deep-water magnetic investigations, where cesium magnetometry is more commonly employed.

Sub-Bottom Profiling

Sub-bottom profiling (SBP) has quickly become one of the primary techniques for seismic subsurface data collection of submerged paleolandscapes and structures. There are a number of different SBPs used for maritime archaeological surveys, with some being simultaneously used aboard a vessel to corroborate more comprehensive data in shallow and deep-water expeditions. These include chirp, boomers, sparkers, pingers, and parametric sonar SBPs, which send out acoustic sound waves to map the seafloor subsurface. A number of these acoustic waves are reflected by the seabed, while others penetrate areas of the seafloor, reaching different sediment horizons. When a buried structure or landscape feature exerts pressure on surrounding sediments, the pressure at a submerged depth rearranges sediment particles creating distinct tightly-packed horizons through compaction and cementation over long periods of time. The original sediments formed at or near the surface become covered by newly deposited material creating buried strata. As acoustic waves encounter different densities and compositions of sediment horizons, they reverberate. A detector, called a hydrophone, counts the time it takes for these signals to return to the vessel in order to identify reflections in sediment and estimate the depth of these horizons.

This data can be used to interpret the location and extent of submerged paleolandscape landforms or geological features, indicating past human activity that may have been subjected to a rise in sea level. SBPs can also help maritime archaeologists approach submerged wreckage that has been partially or completely buried, whether in a pre-contact or historic context. The data is represented as a vertical slice or cross-section graph, which displays a distribution of the different sediment types encountered along with any presence of buried objects and the geological structure below the seafloor. These graphs are typically used to map the thickness of sediment layers and geological depositions based on the data collected. While there are a multitude of different model sub-bottom profilers, this section primarily focuses on some of the most commonly used for maritime archaeological and submerged geophysical inquiry.

A chirp device uses an acoustic swept-frequency signal to create 2D images of upper sediment horizons. These devices are used to detect fine-grained sediments and smaller localized submerged subsurface features with high resolution. Chirp SBPs have difficulties penetrating signals through denser and rockier sediments; however, this can create return signals with a stronger reflection aspect at rock and object surfaces. The system is typically deployed while towed behind or below a vessel with the transducer, which emits pulses of high-frequency acoustic signals.

Boomers are a type of SBP also towed behind a vessel, and use a high-energy discharge to produce acoustic sound waves. They are often deployed in shallow water surveys, but are capable of penetrating up to 100 meters into the seafloor to take 2D images of subsurface features and sediment profiles. They typically use an induction coil and a metal plate to generate an electrical pulse, which sends high-frequency acoustic signals downward for shorter-range high-resolution imaging. Depending on the types of sediment encountered, these devices can achieve resolutions in the range of centimeter accuracy. Similar to boomer SBPs in capabilities of producing high-resolution images, sparkers differ in their production of acoustic signals. They generate high-energy sparks that vaporize seawater, sending strong low-frequency acoustic signals to the seafloor, allowing for deeper penetration of the subsurface. However, this lower frequency sound wave can sometimes not produce as high-resolution 2D images as a boomer system.

Pingers are designed specifically for shallow water surveying, emitting short high-frequency singular pulse waves between 2kHz and 20kHz. While they are capable of 2D imaging in higher resolutions, they are limited in penetration depth and are used primarily when high-resolution imaging is needed in shallower water. When more precise and detailed imaging is required of specific paleolandscape features or objects in thinner sediments of shallow water, parametric SBPs can reduce the interference of background noise in the water. This is done by using two higher frequencies simultaneously that interact with each other to produce narrow low-frequency signals underwater. Parametric SBPs can penetrate up to 50 meters into the subsurface and provide sub-meter resolution of geological layering.

Sub-bottom profiling has become a relatively cost-effective and non-invasive technique that has found its way into submerged archaeological geophysics. Not only has it allowed geoarchaeologists to map the seafloor for GIS reconstruction, but it has rationalized the approaches to investigate and preserve submerged cultural resources. Meticulous digital dissection of the data provided by these instruments allows geoarchaeologists to reconstruct ancient landscapes, wreckage, and buried features with clearer representations. This ultimately creates more effective and safe approaches for divers to collect evidence from submerged archaeological sites when contact with the site is required. It also culminates data from these sites for laboratory research, providing interpretations of sedimentological and geological contexts associated with submerged cultural materials.

Multibeam Echosounder

While sub-bottom profiling produces 2D images of subsurface landscapes and features, multibeam echosounders (MBES) are designed to create 3D maps of the seafloor. They use much higher frequency acoustic signals to collect data about bathymetry, geological formations of the seabed, and submerged structures or wreckage. The acoustic frequencies emitted span a wide area of the seabed to generate detailed 3D maps that are color-coded based on depth. The detection system works similarly to sub-bottom profilers, where the time between the initial pulse of a signal and its return determines depth. More encompassing MBES systems may also collect backscatter data, where the strength of a return signal dictates the reflectivity of the seafloor due to differences in sediment characteristics. The integration of MBES in submerged geoarchaeology has provided archaeologists with a better understanding of past seabed environments around submerged sites, which incorporates 3D imaging to produce more precise geospatial models.

Side-Scan Sonar

Also capable of mapping bathymetry, side-scan sonar produces high-resolution live feed data along a track of the seafloor. Limited by its area of coverage, this type of acoustic sonar is useful for locating submerged objects, structures, geological features, and wreckage, particularly in shallow water. Side-scan sonar transducers are mounted to a towfish behind the vessel, with multiple transducers attached just below the surface of the water. The screen used to program the system is integrated onboard. It can toggle the transducers between low and high frequencies, with lower frequency signals generating wider beams that cover larger areas with less detail and depth penetration. Higher frequencies produce a narrower acoustic beam to take higher-quality images at greater depths.

Traditional 2D sonar imaging has been improved upon in the modern era of maritime studies to include CHIRP (compressed high-intensity radar pulse) integrated into these systems. It continuously emits a sweep of varying frequencies to provide clear, high-resolution 2D imaging of the water column and seafloor. While both down-imaging and side-imaging sonar use CHIRP, side-scan sonar is more commonly used for maritime archaeological survey when paired with geospatial software. Down-imaging provides a detailed view directly below a vessel, while side-scan covers a much broader scope of the seafloor area. GIS applications that allow anomalies to be plotted in real time can simultaneously be used onboard to reexamine specific instances once surveying is complete.

Dual-Beam Sonar

Also known as multibeam sonar, these imaging systems use two high-frequency wide beams to send acoustic signals for more accurate depth and bathymetric measurements. Transducers can be swapped on these systems to include high-resolution CHIRP systems under different models. They are often mounted to the side of small vessels or stationary platforms instead of being towed, and offer less complex programming options than side-scan sonar transducers. Integrated screens range from four to nine inches and provide a suite of tools to reduce background noise, adjust beam angle or frequency, and filter CHIRP color grades.

Dual-beam sonar offers maritime archaeologists the capabilities of broad range high-resolution sonar for small-scale bathymetric surveys. CHIRP-enabled transducers for dual-beam sonar greatly enhance the capabilities of these models. They provide nearly photographic resolution imaging for greater clarity when locating wreckage, seafloor features, objects, and submerged structures in some models. It uses high-frequency narrow-sweeping beams to generate snapshots of an area as the transducer moves along the seabed. These images are projected clearly enough to view aquatic life and the foliage surrounding an area of interest.

Tidal Chemistry

Developing strategies for preservation is a significant point of direction before approaching a submerged archaeological site with direct contact. It includes both geophysical maps and models to represent different aspects of a site, along with knowledge of fluctuations in tidal chemistry to assess the stability of submerged cultural material. In particular, the effects of salinated water can have dramatic corrosive effects on metallic remains when removed from their preserved environments. Geochemical analysis of sediment and water samples can provide long-term data for maritime investigations that periodically undergo several phases of examination. Tidal chemistry plays an important role in determining the oxygenation of submerged environments, which is directly associated with the rate of chemical reactions occurring with depositional sediments and organic matter. While geophysical instruments can aid in understanding the depth and thickness of these sediments, the analysis of these reactions within the site correlates with the conservation of its associated archaeological remains. This creates common ground for maritime archaeology to benefit directly from the fields of chemistry, biology, and geophysics to solve questions about the processes that transform an archaeological site over thousands of years.

Many of the processes aforementioned in the previous section on archaeometry and geochemical analysis are employed when analyzing sediment and solution samples. These include particle size distribution, total dissolved solid testing, refractometry, micromorphology, flotation sampling, spectrophotometry, mass and Fourier transform infrared spectroscopy, along with a plethora of X-ray wavelength techniques. Geochemical analytic methods provide maritime archaeologists with detailed compositional data about chemical bonding, structure, and light absorption factors in different sediments to analyze their reactions to the submerged environment.

With additional insight into biological processes occurring at the site through compound light and SEM microscopy, the distribution of minerals and chemical reactions occurring at a particular site can be examined. As tidal conditions change, the rate of organic decay, salt concentration, and alterations to compounds in the sediment of a site affect the microbiome present. With various types of bacteria and microorganisms involved in such chemical processes, understanding tidal chemistry is paramount to protecting and preserving cultural resources during archaeological examinations.

Sulfate-reducing bacteria (SRB) play a major role in the degradation of submerged materials made from metal, and contribute to the presence of sulfide staining on porous artifacts like pottery. *Desulfovibrio desulfuricans* is a species of gram-negative sulfate-reducing bacteria that is commonly found in salt and freshwater environments where decaying organic matter consumes oxygen, creating localized anaerobic environments. It is typically observed in estuaries and denser marsh areas during low tide, where organic deposits are rich with pockets of microblooms. Sea water flowing upstream brings a large supply of sulfates, and under these pre-established localized anaerobic conditions, sulfate-reducing bacteria use hydrogen as an electron donor to reduce sulfates $(SO_4)^{2-}$. These sulfates act as terminal electron acceptors to reduce to sulfides (S^{2-}) and hydrogen sulfide (H_2S) as a metabolic by-product in a dissimilatory sulfate reduction reaction:

$H_2SO_4 + 8H \rightarrow H_2S + 4H_2O$

This causes discoloration, sulfide staining, and extensive degradation of submerged cultural material that is susceptible to absorption of aqueous solutions, such as pottery. It also imposes significant corrosive action on ferric materials such as submerged iron, causing widespread pitting of these objects due to the production of metal sulfides.

Due to the environmental stress associated with low-oxygenated anaerobic environments, including some wetlands and coastal sites, gram-positive bacteria are not as prevalent as gram-negative species such as *Desulfovibrio desulfuricans*. Gram-positive bacteria have a thicker peptidoglycan cell wall that can act as a barrier against sulfate-reducing bacteria, making it difficult for them to leach nutrients from gram-positive species. Gram-negative bacteria exhibit a thinner peptidoglycan layer, making their protective membranes more permeable.

When encountering species of sulfate-reducing bacteria within sediment or solution samples during microscopic analysis, determining their reaction process with artifacts can provide a basis for a range of conservation techniques. These techniques include the introduction of gram-positive species of bacteria to stabilize and inhibit further corrosive or sulfide-producing reactions due to their thicker cell wall. This can help stabilize the microbiome of materials removed from submerged environments until further analysis and chemical consolidation of the object is required. They are also susceptible to cell denaturing from UVC light ranging between 222 and 254 nanometers, although ranges closer to the upper levels can cause DNA damage to human skin. While there are distinctions between subspecies of SRB that can thrive in alkaline or acidic environments, the optimal pH for this microorganism is generally between 6 and 8.

Among the species of bacteria that affect archaeological remains in submerged anaerobic conditions, sulfate-reducing bacteria are one of the most destructive in creating corrosion and sulfide-producing reactions. There are several other microbes that affect low-oxygenated marine environments but do not exhibit the same sulfide production as SRB. Iron-oxidizing species of bacteria, such as *Leptothrix* and *Acidithiobacillus*, can contribute to the formation of iron oxides by reducing ferrous iron (Fe^{2+}) compounds to ferric iron (Fe^{3+}), which can accelerate corrosion processes. In turn, iron-reducing bacteria (IRB) can reduce the ferric iron present to ferrous iron under anaerobic conditions, which can also impact corrosion reactions. IRB is known to create biofilms on metal surfaces, particularly in steel.

Whether participating in laboratory analysis or engaging in fieldwork, the basic principles of tidal chemistry are the crux of many issues associated with the preservation and data collection of submerged cultural material. Collection of samples in the field can provide more than just the determination of sediment compositions. It reveals the biological processes washing in and out of submerged sites over long periods of time, impacting the overall geological environments in which we find archaeological remains. While this guide does not fully encompass the broader aspects of chemical consolidation for affected materials, the detection methods and salience of these issues while conducting maritime research is a valuable perspective to better preserve and protect submerged cultural resources.

AFTERWORD:
A TRIAL BY SOIL

In any archaeological investigation, whether on land or underwater, the foundations of geoscience play a critical role in disentangling the geological contexts in which we find these sites. This book serves as a field reference for the Geologic Time Scale and archaeological periods, encompassing both hominin evolution and human prehistory to further investigate the means by which technology has aided our understanding of paleoenvironments. As many archaeologists progress through their careers, whether breaking ground for the first time or with decades of experience serving the cultural heritage field, it is understandably easy to overlook the intricate chronology of the various eons, eras, periods, epochs, and ages that broaden the temporal contexts behind our prehistory. Kindling this knowledge can not only deepen our understanding of prehistoric timelines but also bring to light the inadvertent symmetries between the multiple fields of scientific inquiry that underlie archaeology.

With the introduction of geophysical and geospatial technology to both terrestrial and submerged archaeological methods, the necessity to integrate the earth sciences, fundamental physics, and biological studies has pushed the innovative techniques of geoarchaeology into the scope of cultural resource management. More critically, the implementation of this technology, especially GIS software, has aided our interpretations of landscapes relative to the context we expect when conducting fieldwork. By incorporating subsets of data from geophysical surveys into GIS probability-based models, computer software can derive more accurate and precise algorithms to determine the locations of high or low-probability buffer zones in predictive models. As a result, this has streamlined project data collection, which creates more detailed information about landforms when representing the data in 2D or 3D models.

In maritime archaeology, geophysics has had a profound impact on fieldwork and the post-processing of collected data. From the way submerged subsurface data is collected and interpreted to bathymetric seafloor mapping, geophysical technology has fundamentally transformed maritime research. Geoarchaeology offers a variety of complex tools and interdisciplinary knowledge to better dissect paleoenvironments and submerged site data using geophysical and geochemical analytic methods. This elaborate subfield, which combines geological and archaeological studies, has yielded a remarkable symmetry between the earth sciences, physics, biology, and maritime research. Its promising future for developing considerable technological innovations in archaeology warrants its own characterization as a dynamic and complex approach to the field that requires a significant amount of specialization.

REFERENCES

Adovasio, James M., J. D. Gunn, J. Donahue, and R. Stuckenrath. 1978. "Meadowcroft Rockshelter, 1977: An Overview." *American Antiquity* 43, no. 4: 632-651.

Aiello, Leslie C. 2010. "Five years of *Homo floresiensis*." *American Journal of Physical Anthropology* 142, no. 2: 167-79.

Alegret, Laia, Gabriela J. Arreguín-Rodríguez, Carlos A. Trasviña-Moreno, and Ellen Thomas. 2021. "Turnover and stability in the deep sea: Benthic foraminifera as tracers of Paleogene global change." *Global and Planetary Change* 196: 103372.

Alekseyev, Valery Pavlovich. 1986. "The Origin of the Human Race." Moscow: *Progress Publishers.*

Aldeias, Vera, Paul Goldberg, Dennis Sandgathe, Francesco Berna, Harold L. Dibble, Shannon P. McPherron, and Alain Turq et al. "Evidence for Neanderthal use of fire at Roc de Marsal (France)." 2012. *Journal of Archaeological Science* 39, no. 7: 2414-2423.

Alemseged, Zeresenay, Fred Spoor, William H. Kimbel, René Bobe, Denis Geraads, Denné Reed, and Jonathan G. Wynn. 2006. "A juvenile early hominin skeleton from Dikika, Ethiopia." *Nature* 443: 296-301.

Almécija, Sergio, Melissa Tallman, David M. Alba, Marta Pina, Salvador Moya-Sola, and William L. Jungers. 2013. "The femur of Orrorin tugenensis exhibits morphometric affinities with both Miocene apes and later hominins." *Nature Communications* 4, no. 1: 2888.

Almécija, Sergio, Ian J. Wallace, Stefan Judex, David M. Alba, and Salvador Moyà-Solà. 2015. "Comment on "Human-like hand use in *Australopithecus africanus.*" *Science* 348, no. 6239: 1101.

Alves, Eduardo Q., Kita Macario, Philippa Ascough, and Christopher Bronk Ramsey. 2018. "The Worldwide Marine Radiocarbon Reservoir Effect: Definitions, Mechanisms, and Prospects." *Reviews of Geophysics* 56, no. 1: 278-305.

Ao, Hong, Chun-Ru Liu, Andrew P. Roberts, Peng Zhang, and Xinwen Xu. 2017. "An updated age for the Xujiayao hominin from the Nihewan Basin, North China: Implications for Middle Pleistocene human evolution in East Africa." *Journal of Human Evolution* 106: 54-65.

Arambourg, Camille and Yves Coppens. 1967. "On the discovery, in the Lower Pleistocene of the Omo Valley (Ethiopia) of an Australopithecine mandible." *Comptes Rendus de l'Académie des Sciences de Paris, série D* 265L 589-590.

Artioli, Gilberto, Ivana Angelini, Günther Kaufmann, Caterina Canovaro, Gregorio Dal Sasso, and Igor Maria Villa. 2017. "Long-distance connections in the Copper

Age: New Evidence from the Alpine Iceman's copper axe." *PLoS ONE* 12, no. 12: e0189561.

Asfaw, Berhane, Tim D. White, O. Lovejoy, B. Latimer, S. Simpson, and Gen Suwa. 1999. "*Australopithecus garhi*: a new species of early hominid from Ethiopia." *Science* 284, no. 5414: 629-35.

Ashton, Nick, Simon G. Lewis, Isabelle De Groote, Sarah M. Duffy, Martin Bates, Richard Bates, and Peter Hoare. 2014. "Hominin Footprints from Early Pleistocene Deposits Happisburgh, UK." *PLoS ONE* 9, no. 2: e88329.

Bae, Christopher J. and Xiujie Wu. "Making sense of eastern Asian Late Quaternary hominin variability." *Nature* 15, no. 9479.

Backwell, Lucinda and Francesco d'Errico. 2008. "Early hominid bone tools from Drimolen, South Africa." *Journal of Archaeological Science* 35, no. 11: 2880-2894.

Balter, Vincent, Janne Blichert-Toft, José Braga, Philippe Telouk, Francis Thackeray, and Francis Albarède. 2008. "U-Pb dating of fossil enamel from the Swartkrans Pleistocene hominid site, South Africa." *Earth and Planetary Science Letters* 267, no. 1-2: 236-246.

Barat, Flore, Bernard Mercier de Lépinay, Marc Sosson, Carla Müller, Peter O. Baumgartner, and Claudia Baumgartner-Mora. 2014. "Transition from the Farallon Plate subduction to the collision between South and Central America: Geological evolution of the Panama Isthmus." *Tectonophysics* 622: 145-167.

Baumgartner, Raphael J., Martin J. Van Kranendonk, David Wacey, Marco L. Fiorentini, Martin Saunders, Stefano Caruso, and Anais Pages et al. 2019. "Nano-porous pyrite and organic matter in 3.5-billion-year-old stromatolites record primordial life." *Geology* 47, no. 11: 1039-1043.

Becerra-Valdivia, Lorena, Michael R. Waters, Thomas W. Stafford Jr., Sarah L. Anzick, Daniel Comeskey, Thibaut Devièse, and Thomas Higham. 2018. "Reassessing the chronology of the archaeological site of Anzick." *PNAS* 115, no. 27: 7000-7003.

Bennett, Matthew R., David Bustos, Jeffrey S. Pigati, Kathleen B. Springer, Thomas M. Urban, Vance T. Holliday, and Sally C. Reynolds et al. 2021. "Evidence of humans in North America during the Last Glacial Maximum." *Science* 373, no. 6562: 1528-1531.

Benton, Michael J. 2023. "The dinosaur boom in the Cretaceous." *Geological Society* 544: 453-475. London, Special Publications.

Berger, Lee R., Tebogo Vincent Makhubela, Keneiloe Molopyane, Ashley Krüger, Patrick S. Randolph-Quinney, Marina Elliot, and Becca Peixotto. 2023. "Evidence for deliberate burial of the dead by *Homo naledi*." *eLife* 12: RP89106.

Berger, Lee R., John Hawks, Darryl J de Ruiter, Steven Churchill, Peter Schmid, Lucas K Delezene, and Tracy L Kivell et al. 2015. "*Homo naledi*, a new species of the genus *Homo* from the Dinaledi Chamber, South Africa." *eLife* 4: e09560.

Berger, Lee R., Darryl J. de Ruiter, Steven E. Churchill, Peter Schmid, Kristian J. Carlson, Paul H. G. M. Dirks, and Job M. Kibii. 2010. "*Australopithecus sediba*: a new species of Homo-like australopith from South Africa." *Science* 328, no. 5975: 195-204.

Berger, Lee R. 2006. "Brief communication: predatory bird damage to the Taung type-skull of *Australopithecus africanus* Dart 1925." *American Journal of Physical Anthropology* 131, no. 2: 166-8.

Bermúdez de Castro, José María, Eudald Carbonell, Isabel Cáceres, Carlos Diez, Yolanda Fernández-Jalvo, and Marina Mosquera et al. 1999. "The TD6 (Aurora Stratum) hominid site. Final remarks and new questions." *Journal of Human Evolution* 37, no. 3-4: 695-700.

Bermúdez de Castro, José María, J. L. Arsuaga, E. Carbonell, A. Rosas, I. Martínez, and M. Mosquera. 1997. "A Hominid from the Lower Pleistocene of Atapuerca, Spain: Possible Ancestor to Neanderthals and Modern Humans." *Science* 276, no. 5317: 1392-1395.

Bermúdez de Castro, José María and M. Martinón-Torres. 2022. "The origins of the *Homo sapiens* lineage: When and where?" *Quaternary International* 634: 1-13.

Bermúdez de Castro, José María, María Martinón-Torres, Aida Gómez-Robles, Leyre Prado-Simón, Laura Martín-Francés, María Lapresa, and Anthony Olejniczak et al. 2011. "Early Pleistocene human mandible from Sima del Elefante (TE) cave site in Sierra de Atapuerca (Spain): a comparative morphological study." *Journal of Human Evolution* 61, no. 1: 12-25.

Berna, Francesco, Paul Goldberg, Liora Kolska Horwitz, James Brink, Sharon Holt, Marion Bamford, and Michael Chazan. 2012. "Microstratigraphic evidence of in situ fire in the Acheulean strata of Wonderwerk Cave, Northern Cape province, South Africa." *PNAS* 109, no. 20: E1215-E1220.

Beuzen-Waller, Tara, Friederike Stock, and Yasuhisa Kondo. 2018. "Geoarchaeology: A toolbox for revealing latent data in sedimentological and archaeological records." *Quaternary International*. 483, no. 30: 1-4.

Bicket, Andrew, Antony Firth, Louise Tizzard, and Jonathan Benjamin. 2014. "Heritage Management and Submerged Prehistory in the United Kingdom." In *Prehistoric Archaeology on the Continental Shelf*, edited by Amanda M. Evans, Joseph C. Flatman, and Nicholas C. Flemming, 224-225, Springer, 2014.

Bininda-Emonds, Olaf R. P. and Andy Purvis. 2012. "Comment on "Impacts of the Cretaceous Terrestrial Revolution and KPg Extinction on Mammal Diversification." *Science* 337, no. 6090: 34.

Black, Davidson. 1927a. "Further Hominid Remains of Lower Quaternary Age from the Chou Kou Tien Deposit." *Nature* 120, no. 954.

Black, Davidson. 1927b. "On a Lower Molar Hominid Tooth from the Chou Kou Tien Deposit." *Palaeontologica Sinica, New Series D*, no. 7: 1-28.

Black, Davidson. 1929. "*Sinanthropus pekinensis*: The Recovery of Further Fossil Remains of this Early Hominid from the Chou Kou Tien Deposit." *Science* 69, no. 1800: 674-676.

Blain, Hugues-Alexandre, Salvador Bailon, Gloria Cuenca-Bescós, Maria Bennásar, Juan Rofes, Juan Manuel López-García, and Rosa Huguet et al. 2010. "Climate and environment of the earliest West European hominins inferred from amphibian and squamate reptile assemblages: Sima del Elefante Lower Red Unit, Atapuerca, Spain." *Quaternary Science Reviews* 29, no. 23-24: 3034-3044.

Blumenschine, Robert J., Charles R. Peters, Fidelis T. Masao, Ronald J. Clarke, Alan L. Deino, Richard L. Hay, and Carl C. Swisher et al. 2003. "Late Pliocene *Homo* and Hominid Land Use from Western Olduvai Gorge, Tanzania." *Science* 299, no. 5610: 1217-1221.

Bobe, René. 2006. "The evolution of arid ecosystems in eastern Africa." *Journal of Arid Environments* 66, no. 3: 564-584.

Bobe, René, Frederick Kyalo Manthi, Carol V. Ward, J. Michael Plavcan, and Susana Carvalho. 2020. "The ecology of *Australopithecus anamensis* in the early Pliocene of Kanapoi, Kenya." *Journal of Human Evolution* 140, no. 102717.

Boisserie, Jean-Renaud. 2010. "Ardipithecus ramidus and the Birth of Humanity." *Annales d'Éthiopie* 25, no. 1: 271-281.

Boldurian, Anthony. 2008. "Clovis Type-Site, Blackwater Draw, New Mexico: A History, 1929-2009." *North American Archaeologist* 29, no. 1: 70.

Boltwood, Bertram B. 1907. "Ultimate Disintegration Products of the Radio-Active Elements, Part II: Disintegration Products of Uranium." *American Journal of Science* 23: 77-88.

Bond, David P. G. and Stephen E. Grasby. 2020. "Late Ordovician mass extinction caused by volcanism, warming, and anoxia, not cooling and glaciation." *Geology* 48, no. 8: 777-781.

Borrell, Ferran. 2017. "Bidirectional blade technology in the Near East during the Pre-Pottery Neolithic B [La technologie de débitage laminaire bipolaire au Proche-Orient durant le Néolithique précéramique B (PPNB)]." *Journal of Lithic Studies* 4, no. 2: 129-161.

Botha, J., F. Abdala, and R. Smith. 2007. "The oldest cynodont: new clues on the origin and early diversification of the Cynodontia." *Zoological Journal of the Linnean Society* 149, no. 3: 477-492.

Boyadzhiev, Kamen. 2018. "Warfare in the Chalcolithic of Bulgaria." In *Band 293 Southeast Europe and Anatolia in prehistory. Essays in honor of Vassil Nikolov on his 65th anniversary*: 263-264. Verlag Dr. Rudolf Hablet (2016).

Braga, José, G. Chinamatira, B. Zipfel, and V. Zimmer. 2022. "New fossils from Kromdraai and Drimolen, South Africa, and their distinctiveness among *Paranthropus robustus*." *Scientific Reports* 12: 13956.

Braun, David R., Vera Aldeias, Will Archer, J. Ramon Arrowsmith, Niguss Baraki, Christopher J. Campisano, and Alan L. Deino et al. 2019. "Earliest known

Oldowan artifacts at >2.58 Ma from Ledi-Geraru, Ethiopia, highlight early technological diversity." *PNAS* 116, no. 24: 11712-11717.

Brock, A., and G. Ll. Isaac. 1974. "Paleomagnetic stratigraphy and chronology of hominid-bearing sediments east of Lake Rudolf, Kenya." *Nature* 247: 344-348.

Broom, R. 1938. "The Pleistocene Anthropoid Apes of South Africa." *Nature* 142: 377-379.

Brown, F. H., and Barbara P. Nash. 2014. "Correlation: Volcanic Ash, Obsidian." in *Treatise of Geochemistry* 14.6, no. 2: 63-80. Elsevier.

Brown, F., J. Harris, R. Leakey, and A. Walker. 1985. "Early Homo erectus skeleton from west Lake Turkana, Kenya." *Nature* 316, no. 6031: 788-92.

Brown, Peter. 2012. "LB1 and LB6 *Homo floresiensis* are not modern human (*Homo sapiens*) cretins." *Journal of Human Evolution* 62, no. 2: 201-224.

Brown, Peter, and Tomoko Maeda. 2009. "Liang Bua *Homo floresiensis* mandibles and mandibular teeth: a contribution to the comparative morphology of a new hominin species." *Journal of Human Evolution* 57, no. 5: 571-96.

Brown, Peter, T. Sutikna, M. J. Morwood, R. P. Soejono, Jatmiko, E. Wayhu Saptomo, and Rokis Awe Due. 2004. "A new small-bodied hominin from the late Late Pleistocene of Flores, Indonesia." *Nature* 431: 1055-1061.

Brown, Samantha, Diyendo Massilani, Maxim B. Kozlikin, Michael V. Shunkov, Anatoly P. Derevianko, Alexander

Stoessel, and Blair Jope-Street et al. 2022. "The earliest Denisovans and their cultural adaptation." *Nat Ecol Evol.* 6, no. 1: 28-35.

Brown, Samantha, Naigui Wang, Annette Oertle, Maxim B. Kozlikin, Michael V. Shunkov, Anatoly P. Derevianko, and Daniel Comeskey et al. 2021. "Zooarchaeology through the lens of collagen fingerprinting at Denisova Cave." *Scientific Reports* 11, no. 15457.

Brumm, Adam, Gerrit D. van den Bergh, Michael Storey, Iwan Kurniawan, Brent V. Alloway, Ruly Setiawan, and Erick Setiyabudi et al. 2016. "Age and context of the oldest known hominin fossils from Flores." *Nature* 534: 249-253.

Brumm, Adam, Mark, W. Moore, Gert D. van den Bergh, Iwan Kurniawan, Michael J. Morwood, and Fachroel Aziz. 2010. "Stone technology at the Middle Pleistocene site of Mata Menge, Flores, Indonesia." *Journal of Archaeological Science* 37, no. 3: 451-473.

Bruner, Emiliano and Marlize Lombard. 2020. "The skull from Florisbad: a paleoneurological report." *Journal of Anthropological Science* 98.

Brunet, Michel, Franck Guy, David Pilbeam, Hassane Taisso Mackaye, Andossa Likius, Djimdoumalbaye Ahounta, and Alain Beauvilain et al. 2002. "A new hominid from the Upper Miocene of Chad, Central Africa." *Nature* 418: 145-151.

Burek, Cynthia V. 2001. "The first lady geologist, or collector par excellence?" *Geology Today* 17, no. 5: 1-8.

Bushman, Jess R. 1983. "Hutton's Uniformitarianism." *Brigham Young University Studies* 23, no. 1: 41-48.

Cabreira, Sergio F., Cesar L. Schultz, Lúcio R. da Silva, Luiz Henrique Puricelli Lora, Cristiane Pakulski, Rodrigo C. B. do Rêgo, and Marina B. Soares et al. 2022. "Diphyodont tooth replacement of *Brasilodon* - A Late Triassic eucynodont that challenges the time of origin of mammals." *Journal of Anatomy* 241, no. 6: 1424-1440.

Campaña, I., A. Pérez-González, A. Benito-Calvo, J. Rosell, R. Blasco, J. M. Bermúdez de Castro, and Eudald Carbonell et al. 2016. "New interpretation of the Gran Dolina-TD6 bearing *Homo antecessor* deposits through sedimentological analysis." *Scientific Reports* 6, no. 34799.

Caan, Rebecca L., Mark Stoneking, and Allan C. Wilson. 1987. "Mitochondrial DNA and human evolution." *Nature* 325: 31-36.

Capel, Eliott, Christopher J. Cleal, Jinzhuang Xue, Claude Monnet, Thomas Servais, and Borja Cascales-Miñana. 2022. "The Silurian-Devonian terrestrial revolution: Diversity patterns and sampling bias of the vascular plant macrofossil record." *Earth-Science Reviews* 231: 104085.

Cerling, Thure E., B. W. Cerling, G. H. Curtis, R. E. Drake, and F. H. Brown. 1979. "Preliminary correlations between the Koobi Fora and Shungura Formations, East Africa." *Nature* 279: 118-121.

Cerling, Thure E., Naomi E. Levin, Jay Quade, Jonathan G. Wynn, David L. Fox, John D. Kingston, and Richard G. Klein et al. 2010. "Comment on the Paleoenvironment of *Ardipithecus ramidus*." *Science* 328, no. 5982: 1105.

Chen, Fahu, Frido Welker, Chuan-Chou Shen, Shara E. Bailey, Inga Bergmann, Simon Davis, and Huan Xia et al. 2019. "A late Middle Pleistocene Denisovan mandible from the Tibetan Plateau." *Nature* 569: 409-412.

Chiarenza, Alfio Alessandro, Alexander Farnsworth, Philip D. Mannion, Daniel J. Lunt, Paul J. Valdes, Joanna V. Morgan, and Peter A. Allison. 2020. "Asteroid impact, not volcanism, caused the end-Cretaceous dinosaur extinction." *PNAS* 117, no. 29: 17084-17093.

Childe, V. Gordan. 1936. *Man Makes Himself*. London: Watts & Co., 66-104

Ciborowski, T. Jake R., David J. Nash, Timothy Darvill, Ben Chan, Mike Parker Pearson, Rebecca Pullen, and Colin Richards et al. 2024. "Local and exotic sources of sarsen debitage at Stonehenge revealed by geochemical provenancing." *Journal of Archaeological Science: Reports* 53, 104406.

Clark, Jorie, Anders E. Carlson, Alberto V. Reyes, Elizabeth C. B. Carlson, Louise Guillaume, Glenn A. Milne, and Lev Tarasov et al. 2022. "The age of the opening of the Ice-Free Corridor and implications for the peopling of the Americas." *Proc. Natl. Acad. Sci. U.S.A* 119, no. 14: e2118558119.

Cline, Eric H. 2014. *1177 B.C.: The Year Civilization Collapsed*. Princeton University Press.

Cortell-Nicolau, Alfredo, Alastair Key, and Antonio Palomo. 2025. "Assessing geometric microliths as cultural markers through an analysis of shape variation and projectile performance." *Scientific Reports* 15, no. 11247.

Coté, Mark. 2012. "The Prehistoric Turn?: Networked New Media, Mobility, and the Body." *The International Encyclopedia of Media Studies*, edited by Kelly Gates. Oxford, UK: Blackwell, 2012.

Couvreur, Thomas LP, Gilles Dauby, Anne Blach-Overgaard, Vincent Deblauwe, Steven Dessein, Vincent Droissart, and Oliver J. Hardy et al. 2020. "Tectonics, climate and the diversification of the tropical African terrestrial flora and fauna." *Biol Rev Camb Philos Soc* 96, no. 1: 16-51.

Cross, Timothy and Peter Homewood. 1997. "Amanz Gressly's role in founding modern stratigraphy." *Geological Society of America Bulletin* 109, no. 12: 1617-1630.

Curtis, G. H., Drake, T. Cerling, and Hampel. 1975. "Age of KBS Tuff in Koobi Fora Formation, East Rudolf, Kenya." *Nature* 258: 395-398.

Dart, Raymond A. 1925. "Australopithecus africanus The Man-Ape of South Africa." *Nature* 115: 195-199.

Daver, G., F. Guy, H. T. Mackaye, A. Likius, J. -R. Boisserie, A. Moussa, and L. Pallas et al. 2022. "Postcranial evidence of late Miocene hominin bipedalism in Chad." *Nature* 609: 94-100.

Davies, J. H. F. L., A. Marzoli, H. Bertrand, N. Youbi, M. Ernesto, and U. Schaltegger. 2017. "End-Triassic mass extinction started by intrusive CAMP activity." *Nature Communications* 8, no. 15596.

Day, M. H., and E. H. Wickens. 1980. "Laetoli Pliocene hominid footprints and bipedalism." *Nature* 286: 385-387.

Dean, Christopher D., Alfio Alessandro Chiarenza, Jeffrey W. Doser, Alexander Farnsworth, Lewis A. Jones, Sinéad J. Lyster, and Charlotte L. Outhwaite et al. 2025. "The structure of the end-Cretaceous dinosaur fossil record in North America." *Current Biology* 35, no. 9: 1973-1988.e6.

De la Vega, Elwyn, Thomas B. Chalk, Paul A. Wilson, Ratna Priya Bysani, and Gavin L. Foster. 2020. "Atmospheric □□$_2$ during the Mid-Piacenzian Warm Period and the M2 glaciation." *Scientific Reports* 10, no. 11002.

Delson, Eric and Chris Stringer. 2022. "The naming of *Homo bodoensis* by Roksandic and colleagues does not resolve issues surrounding Middle Pleistocene human evolution." *Evolutionary Anthropology* 31: 233-236.

Demuro, Martina, Lee J. Arnold, Josep-María Parés, Arantza Aranburu, Rosa Huguet, Josep Vallverdú, and Juan-Luis Arsuaga et al. 2022. *Quaternary Geochronology* 71, no. 101318.

Dennell, Robin W., Julien Louys, Hannah J. O'Regan, and David M. Wilkinson. 2014. "The origins and persistence of *Homo floresiensis* on Flores: biogeographical and ecological perspectives." *Quaternary Science Reviews* 96, no. 15: 98-107.

Derevianko, A. P., Michael V. Shunkov, and Pavel Valerievich Volkov. 2008. "A Paleolithic Bracelet from Denisova Cave." *Archaeology Ethnology and Anthropology of Eurasia* 34, no. 2: 13-25.

De Ruiter, Darryl J., Myra F. Laird, Marina Elliot, Peter Schmid, Juliet Brophy, John Hawks, and Lee R. Berger. 2019. "*Homo naledi* cranial remains from the Lesedi chamber of the rising star cave system, South Africa." *Paleoanthropology* 2019: 1-26.

Deter-Wolf, Aaron, Benoît Robitaille, Lars Krutak, and Sébastien Galliot. 2016. "The world's oldest tattoos." *Journal of Archaeological Science: Reports* 5: 19-24.

Deter-Wolf, Aaron, Benoît Robitaille, Danny Riday, Aurélien Burlot, and Maya Sialuk Jacobsen. 2024. "Chalcolithic Tattooing: Historical and Experimental Evaluation of the Tyrolean Iceman's Body Markings." *European Journal of Archaeology* 27, no. 3: 267-288.

Détroit, Florent, Armand Salvador Mijares, Julien Corny, Guillaume Daver, Clément Zanolli, Eusebio Dizon, and Emil Robles et al. 2019. "A new species of *Homo* from the Late Pleistocene of the Phillipines." *Nature* 568: 181-186.

Dibble, Harold L., Vera Aldeias, Paul Goldberg, Shannon P. McPherron, Dennis Sandgathe, and Teresa E. Steele. 2015. "A critical look at evidence from La Chapelle-aux-Saints supporting an intentional Neanderthal burial." *Journal of Archaeological Science* 53: 649-657.

Dietrich, Laura, Eva Götting-Martin, Jasmine Hertzog, Philippe Schmitt-Kopplin, Patrick E. McGovern,

Gretchen R. Hall, and Christian W. Peterson et al. 2020. "Investigating the function of Pre-Pottery Neolithic stone troughs from Göbekli Tepe - An integrated approach." *Journal of Archaeological Science: Reports* 34, part A.

Díez, J. Carlos, Yolanda Fernández-Jalvo, Jordi Rosell, Isabel Cáceres. 1999. "Zooarchaeology and taphonomy of Aurora Stratum (Gran Dolina, Sierra de Atapuerca, Spain)." *Journal of Human Evolution* 37, no. 3-4: 623-652.

Dillehay, Tom D. and Michael B. Collins. 1988. "Early cultural evidence from Monte Verde in Chile." *Nature* 332: 150-152.

Dillehay, Tom D., Carlos Ocampo, José Saavedra, Andre Oliveira Sawakuchi, Rodrigo M. Vega, Mario Pino, and Michael B. Collins. 2015. "New Archaeological Evidence for an Early Human Presence at Monte Verde, Chile." *PLOS ONE* 10, no. 12: e0145471.

Dirks, Paul H. G. M., Eric M. Roberts, Hannah Hilbert-Wolf, Jan D. Kramers, John Hawks, Anthony Dosseto, and Mathieu Duval et al. 2017. "The age of *Homo naledi* and associated sediments in the Rising Star Cave, South Africa." *eLife* 6: e242231.

Dirks, Paul H. G. M., Lee R. Berger, Eric M. Roberts, Jan D. Kramers, John Hawks, Patrick S. Randolph-Quinney, and Marina Elliot et al. 2015. "Geological and taphonomic context for the new hominin species *Homo naledi* from the Dinaledi Chamber South Africa." *eLife* 4: e09561.

Dominguez-Rodrigo, Manuel. 2014. "Is the "Savanna Hypothesis" a Dead Concept for Explaining the Emergence of the Earliest Hominins?" *Current Anthropology* 55, no. 1: 59-81.

Doronicheva, E. V., L. V. Golovanova, J. V. Kostina, S. A. Legkov, G. N. Poplevko, E. I. Revina, and O. Y. Rusakova et al. 2022. "Functional characterization of Mousterian tools from the Caucasus using comprehensive use-wear and residue analysis." *Scientific Reports* 12, no. 17421.

Drake, R. E., G. H. Curtis, T. E. Cerling, B. W. Cerling, and J. Hampel. 1980. "KBS Tuff dating and geochronology of tuffaceous sediments in the Koobi Fora and Shungura Formations, East Africa." *Nature* 283: 368-372.

Dubois, Eugene. 1894. "*Pithecanthropus erectus*, a transitional form." New York: G. E. Stechert (Alfred Hafner).

Duval, Mathieu, Rainer Grün, Josep M. Parés, Laura Martín-Francés, Isidoro Campaña, Jordi Rosell, and Qingfeng Shao et al. 2018. "The first direct ESR dating of a hominin tooth from Atapuerca Gran Dolina TD-6 (Spain) supports the antiquity of *Homo antecessor*." *Quaternary Geochronology* 47: 120-137.

Edinburgh Geological Society. 2020. "Sir Charles Lyell (1797-1875)." *Edinburghgeolsoc.org*.

Eldrett, James S., David R. Greenwood, Ian C. Harding, and Matthew Huber. 2009. "Increased seasonality through the Eocene to Oligocene transition in northern high latitudes." *Nature* 459, no. 7249: 969-73.

Ell, T. 2011. "Two Letters of Signor Giovanni Arduino, Concerning His Natural Observations: First Full English Translation. Part 2." Edited by D. Oldroyd. *Earth Sciences History* 30, no. 2: 267-286.

Ell, T. 2012. "Two Letters of Signor Giovanni Arduino, Concerning His Natural Observations: First Full English Translation. Part 2." Edited by D. Oldroyd. *Earth Sciences History* 31, no. 2: 168-192.

El-Qady, Gad, Mohamed Metwaly, and Mahmut Göktuğ Drahor. 2019. "Geophysical Techniques Applied in Archaeology." In *Archaeogeophysics*, edited by Gad El-Qady and Mohamed Metwaly, 3-4. Springer, 2019.

Erlandson, Jon M., Michael H. Graham, Bruce J. Bourque, Debra Corbett, James A. Estes, and Robert S. Steneck. 2007. "The Kelp Highway hypothesis: Marine ecology, the coastal migration theory, and the peopling of the Americas." *Journal of Island and Coastal Archaeology* 2, no. 2: 161-174.

Fernández-Jalvo, Yolanda, J. Carlos Díez, Isabel Cáceres, and Jordi Rosell. "Human cannibalism in the Early Pleistocene of Europe (Gran Dolina, Sierra de Atapuerca, Burgos, Spain)." *Journal of Human Evolution* 37, no. 3-4: 591-622.

Fiedel, Stuart J. 2017. "The Anzick genome proves Clovis is first, after all." *Quaternary International* 444, Part B: 4-9.

Fiedel, Stuart J. 2018. "Origins of the first Americans: Before and after the Anzick genome." *Reviews in Anthropology* 46, no. 4: 164-179.

Fitch, F. J. and J. A. Miller. 1970. "New Hominid Remains and Early Artefacts from Northern Kenya: Radioisotopic Age Determinations of Lake Rudolf Artefact Site." *Nature* 226: 226-228.

Fitch, F. J., P. J. Hooker, and J. A. Miller. 1976. "40Ar/39Ar dating of the KBS Tuff in Koobi Fora Formation, East Rudolf, Kenya." *Nature* 263: 740-744.

Fladmark, Knut R. 1979. "Routes: Alternate Migration Corridors for Early Man in North America." *American Antiquity* 44, no. 1: 55-69.

Flynn, Andrew G., Stephen L. Brusatte, Alfio Alessandro Chiarenza, Jorge García-Girón, Adam J. Davis, C. Will Fenley IV, and Caitlin E. Leslie et al. 2025. "Late-surviving New Mexican dinosaurs illuminate high end-Cretaceous diversity and provinciality." *Science* 390, no. 6771: 400-404.

Foecke, Kimberly K., Alain Queffelec, Robyn Pickering, and Katerina Harvati. 2024. "No Sedimentological Evidence for Deliberate Burial by *Homo naledi* - A Case Study Highlighting the Need for Best Practices in Geochemical Studies Within Archaeology and Paleoanthropology." *Paleoanthropology* 10, no. 48738.

Foster, Laura C., Daniela N. Schmidt, Ellen Thomas, Sandra Arndt, and Andy Ridgwell. 2013. "Surviving rapid climate change in the deep sea during the Paleogene hyperthermals." *Proc. Natl. Acad. Sci.* 110, no. 23: 9273-9276.

Fu, Qiaomei, Fan Bai, Huiyun Rao, Shaokun Chen, Yannan Ji, Aoran Liu, and E. Andrew Bennett et al. 2025a. "The proteome of the late Middle Pleistocene Harbin individual." *Science* 0, eadu9677.

Fu, Qiaomei, Peng Cao, Qingyan Dai, E. Andrew Bennett, Xiaotian Feng, Melinda A. Yang, and Wanjing Peng et al. 2025b. "Denisovan mitochondrial DNA from dental calculus of the >146,000-year-old Harbin cranium." *Cell* 188, no. 15: 3919-3926.E9.

Gabunia, L. and A. Vekua. 1995. "A Plio-Pleistocene hominid from Dmanisi, East Georgia, Caucasus." *Nature* 373, no. 6514: 509-12.

Gibbard, P.L. 2019. "Giovanni Arduino - the Man Who Invented the Quaternary." *Quaternary International: The Journal of the International Union for Quaternary Research* 500: 11-19.

Gibbon, Ryan J., Travis Rayne Pickering, Morris B. Sutton, Jason L. Heaton, Kathleen Kuman, Ron J. Clarke, and C. K. Brain et al. "Cosmogenic nuclide burial dating of hominin-bearing Pleistocene cave deposits at Swartkrans, South Africa." *Quaternary Geochronology* 24: 10-15.

Gleadow, A. J. W. 1980. "Fission track age of the KBS Tuff and associated hominid remains in northern Kenya." *Nature* 284: 225-230.

Goldberg, P., S. Weiner, O. Bar-Yosef, Q. Xu, and J. Liu. 2001. "Site formation processes at Zhoukoudian, China." *Journal of Human Evolution* 41, no. 5: 483-530.

Goren-Inbar, Naama, Nira Alperson-Afil, Mordechai E. Kislev, Orit Simchoni, Yoel Melamed, Adi Ben-Nun, and Ella Werker. 2004. "Evidence of Hominin Control of

Fire at Gesher Benot Ya'aqov, Israel." *Science* 304, no. 5671: 725-7.

Goren, Yuval. 2008. "The location of specialized copper production by the lost wax technique in the Chalcolithic Southern Levant." *Geoarchaeology* 23, no. 3: 374-397.

Gosic, Milena and Isaac Gilead. 2014. "Fifty Years Later: A Critical Review of the Stratigraphy, Chronology, and Context of the Nahal Mishmar Hoard." *Journal of the Israel Prehistoric Society* 44: 226-239.

Groves, C. P., and V. Mazák. 1975. "An Approach to the Taxonomy of the Hominidae: Gracile Villafranchian Hominids of Africa. *Journal of Mineralogy and Geology* 20: 225-247.

Grün, Rainer and Chris Stringer. 2023. "Direct dating of human fossils and the ever-changing story of human evolution." *Quaternary Science Reviews* 322, no. 108379.

Grün, Rainer, Pei-Hua Huang, Wanpo Huang, Frank McDermott, Alan Thorne, Chris B. Stringer, and Ge Yan et al. 1998. "ESR and U-series analyses of teeth from the palaeoanthropological site of Hexian Anhui Province, China." *Journal of Human Evolution* 34: 555-564.

Gulick, Sean P. S. 2024. "End of the Cretaceous." *Geological Society* 544: 549-570. London, Special Publications.

Guy, Franck, Hassane-Taïsso Mackaye, Andossa Likius, Patrick Vignaud, Matthieu Schmittbuhl, and Michel Brunet. 2008. "Symphyseal shape variation in extant and fossil hominoids, and the symphysis of *Australopithecus bahrelghazali*." *Journal of Human Evolution* 55, no. 1: 37-47.

Haile-Selassie, Yohannes. 2021. "From Trees to the Ground: The Significance of *Australopithecus anamensis* in Human Evolution." *Journal of Anthropological Research* 77, no. 4.

Haile-Selassie, Yohannes. 2001. "Late Miocene hominids from the Middle Awash, Ethiopia." *Nature* 412, no. 6843: 178-181.

Haile-Selassie, Yohannes, Stephanie M. Melillo, Antonino Vazzana, Stefano Benazzi, and Timothy M. Ryan. 2019. "A 3.8-million-year-old hominin cranium from Woranso-Mille, Ethiopia." *Nature* 573: 214-219.

Haile-Selassie, Yohannes, Gen Suwa, and Tim D. White. 2004. "Late Miocene teeth from Middle Awash, Ethiopia, and early hominid dental evolution." *Science* 303, no. 5663: 1503-5.

Haile-Selassie, Yohannes, Luis Gibert, Stephanie M. Melillo, Timothy M. Ryan, Mulugeta Alene, Alan Deino, and Naomi E. Levin et al. 2015. "New species from Ethiopia further expands Middle Pliocene hominin diversity." *Nature* 521: 483-488.

Halla, Janna, Nora Noffke, Humberto Reis, Stanley Awramik, Andrey Bekker, Alexander Brasier, and Flávia Callefo et al. 2024. "Ratification of the Base of the ICS Geological Time Scale: The Global Standard Stratigraphic Age (GSSA) for the Hadean Lower Boundary." *Journal of International Geoscience* 47, no. 2: 381-389.

Hanon, Raphaël, Francesco d'Errico, Lucinda Backwell, Sandrine Prat, Stéphane Péan, and Marylène Patou-Mathis. 2021. "New evidence of bone tool use by Early

Pleistocene hominins from Cooper's D, Bloubank Valley, South Africa." *Journal of Archaeological Science: Reports* 39: 103129.

Harari, Yuval Noah. 2015. *Sapiens*. Harper.

Harcourt-Smith, William H. E. 2010. "The First Hominins and the Origins of Bipedalism." *Evolution: Education and Outreach* 3: 333-340.

Harmand, Sonia, Jason E. Lewis, Craig S. Feibel, Christopher J. Lepre, Sandrine Prat, Arnaud Lenoble, and Xavier Boës et al. 2015. "3.3-million-year-old stone tools from Lomekwi 3, West Turkana, Kenya." *Nature* 521: 310-315.

Harrison, T. Mark. 2009. "The Hadean Crust: Evidence from >4 Ga Zircons." *Annu. Rev. Earth Planet. Sci* 37: 479-505.

Harvati, Katerina and Hugo Reyes-Centeno. 2022. "Evolution of *Homo* in the Middle and Late Pleistocene." *Journal of Human Evolution* 173, no. 103279.

Haung, Chao, Jingshu Li, and Xing Gao. 2022. "Evidence of Fire Use by *Homo erectus pekinensis*: An XRD Study of Archaeological Bones from Zhoukoudian Locality 1, China." *Frontiers in Earth Sciences* 9, no. 811319.

Hawks, John, Marina Elliot, Peter Schmid, Steven E. Churchill, Darryl J. de Ruiter, Eric M. Roberts, and Hannah Hilbert-Wolf et al. 2017. "New fossil remains of *Homo naledi* from the Lesedi Chamber, South Africa." *eLife* 6: e24232.

Henry, Amanda G. 2017. "Neanderthal Cooking and the Costs of Fire." *Current Anthropology* 58, no. S16.

Herries, Andy I. R. and Justin W. Adams. 2013. "Clarifying the context, dating and age range of the Gondolin hominins and Paranthropus in South Africa." 65, no. 5: 676-81.

Hillhouse, J. W., J. W. M. Ndombi, A. Cox, and A. Brock. 1977. "Additional results on paleomagnetic stratigraphy of the Koobi Fora Formation, east of Lake Turkana (Lake Rudolf), Kenya." *Nature* 265: 411-415.

Hillhouse, J. W., Thure E. Cerling, and F. H. Brown. 1986. "Magnetostratigraphy of the Koobi Fora Formation, Lake Turkana, Kenya." 91(B11): 11581-11595.

Hoebe, P. W., K. M. Cohen, F. S. Busschers, S. van Heteren, and J. H. M. Peeters. 2024. "Early Holocene inundation of Doggerland and its impact on hunter-gatherers: An inundation model and dates-as-data approach." *Quaternary International* 694: 26-50.

Hoffecker, John F. 2018. "The complexity of Neanderthal technology." *PNAS* 115, no. 9: 1959-1961.

Hu, Wangjie, Ziqian Hao, Pengyuan Du, Fabio Di Vincenzo, Giorgio Manzi, Jialong Cui, and Yun-Xin Fu et al. 2023. "Genomic interference of a severe human bottleneck during the Early to Middle Pleistocene transition." *Science* 381, no. 6661: 979-984.

Hublin, Jean-Jacques, Abdelouahed Ben-Ncer, Shara E. Bailey, Sarah E. Friedline, Simon Neubauer, Matthew M. Skinner, and Inga Bergmann et al. 2017. "New fossils from Jebel Irhoud, Morocco and the pan-African origin of *Homo sapiens*." *Nature* 546: 289-292.

Huguet, Rosa, Xosé Pedro Rodríguez-Álvarez, María Martinón-Torres, Josep Vallverdú, Juan Manuel López-García, Marina Lozano, and Marco Terradillos-Bernal et al. 2025. "The earliest human face of Western Europe." *Nature* 640: 707-713.

Hunt, Chris O., Emma Pomeroy, Tim Reynolds, Emily Tilby, and Graeme Barker. 2023. "*Shanidar et ses fleurs?* Reflections on the palynology of the Neanderthal 'Flower Burial' hypothesis." *Journal of Archaeological Science* 159, 105822.

Hurford, Anthony J., A. J. W. Gleadow, and C. W. Naeser. 1976. "Fission-track dating of pumice from the KBS Tuff, East Rudolf, Kenya." *Nature* 263: 738-740.

Hutchinson, David K., Helen K. Coxall, and Daniel J. Lunt et al. 2021. "The Eocene-Oligocene transition: a review of marine and terrestrial proxy data, models and model-data comparisons." *European Geoscience Union* 17: 269-315.

Ingicco, T., G. D. van den Bergh, C. Jago-on, J.-J. Bahain, M. G. Chacón, N. Amano, and H. Forestier et al. 2018. "Earliest known hominin activity in the Philippines by 709 thousand years ago." *Nature* 557: 233-237.

Jacobs, Guy S., Georgi Hudjashov, Lauri Saag, Pradiptajati Kusuma, Chelzie C. Darusallam, Daniel J. Lawson, and Mayukh Mondal et al. 2019. "Multiple Deeply Divergent Denisovan Ancestries in Papuans." *Cell* 177, no. 4: 1010-1021.

Jacobs, Zenobia, Elena Zavala, Bo Li, Kieran O'Gorman, Michael V. Shunkov, Maxim B. Kozlikin, and Anatoly P.

Derevianko et al. 2025. "Pliestocene chronology and history of hominins and fauna at Denisova Cave." *Nature Communications* 16, no. 4738.

Ji, Qiang, Wensheng Wu, Yannan Ji, Qiang Li, and Xijun Ni. 2021. "Late Middle Pleistocene Harbin cranium represents a new *Homo* species." *The Innovation* 2, no. 3: 100132.

Johanson, Donald C., Tim D. White, and Yves Coppens. 1978. "A new species of the genus *Australopithecus* (Primates: Hominidae) from the Pliocene of eastern Africa." *Kirtlandia* 28: 1-14.

Jöns, H., and J. Harff. 2014. "Geoarchaeological Research Strategies in the Baltic Sea Area: Environmental Changes, Shoreline-Displacement and Settlement Strategies." In *Prehistoric Archaeology on the Continental Shelf*, edited by Amanda M. Evans, Joseph C. Flatman, and Nicholas C. Flemming, 178, Springer, 2014.

Kaifu, Yousuke, Iwan Kurniawan, Soichiro Mizushima, Junmei Sawada, Michael Lague, Indra Sutisna, and Unggul P. Wibowo et al. 2024. "Early evolution of small body size in *Homo floresiensis*." *Nature* 15, 6381.

Kalb, Jon E., Craig B. Wood, Charles Smart, Elizabeth B. Oswald, Asseda Mabrete, Sleshi Tebedge, and Paul Whitehead. 1980. "Preliminary geology and paleontology of the Bodo D'ar hominid Site, Afar, Ethiopia." *Palaeogeography, Palaeoclimatology, Palaeoecology* 30: 107-120.

Kaplan, Matt. 2012. "Million-year-old ash hints at origins of cooking." *Nature*

Kaya, Murat and Gül Kaya. 2024. "Mining Processing and Casting in Hittites." *Journal of Anthropological and Archaeological Sciences* 9, no. 2: 1215-1225.

Kender, Sev, Kara Bogus, Gunver K. Pedersen, Karen Dybkjær, Tasmin A. Mather, Erica Mariani, and Andy Ridgwell et al. 2021. "Paleocene/Eocene carbon feedbacks triggered by volcanic activity." *Nature Communications* 12, no. 5186.

Kimbel, William H., Donald C. Johanson, and Yoel Rak. 1998. "Systematic assessment of a maxilla of *Homo* from Hadar, Ethiopia." *American Journal of Physical Anthropology* 103, no. 2: 235-262.

Knoll, AH, E. J. Javaux, D. Hewitt, and P. Cohen. 2006. "Eukaryotic Organisms in Proterozoic Oceans." *Philos Trans R Soc Lond B Biol Sci* 361, no. 1470: 1023-1038.

Ko, Kwang Hyun. 2016. "Hominin interbreeding and the evolution of human variation." *Journal of Biological Research* 23, no. 17.

Koutsoukos, E.A.M. 2005. "Stratigraphy: Evolution of a Concept." In *Applied Stratigraphy. Topics in Geobiology* 23: 3-19. Springer, Dordrecht.

Kröger, Björn, Franziska Franeck, and Christian M. Ø. Rasmussen. 2019. "The Evolutionary Dynamics of the Early Paleozoic Marine Biodiversity Accumulation." *Proc Biol Sci* 286, no. 1909: 20191634.

Kuhlwilm, Martin, Ilan Gronau, Melissa, J. Hubisz, Cesare de Filippo, Javier Prado-Martinez, Martin Kircher, and Qiaomei Fu et al. 2016. "Ancient gene flow from early modern humans into Eastern Neanderthals." *Nature* 530, no. 7591: 429-433.

Lambard, Jean-Baptiste, Alison Pereira, Pierre Voinchet, Hervé Guillou, Marian C. Reyes, Sébastien Nomade, and Xavier Gallet et al. 2024. "Geochronological advances in human and proboscideans first arrival date in the Philippines archipelago (Cagayan valley, Luzon Island)." *Quaternary Geochronology* 84, no. 101597.

Lazaridis, Losif, Alissa Mittnik, Nick Patterson, Swapan Mallick, Nadin Rohland, Saskia Pfrengle, and Anja Furtwängler et al. 2017. "Genetic origins of the Minoans and Mycenaeans." *Nature* 548 (7666): 214-218.

Leakey, L. S. B. 1931. *The Stone-Age Cultures of Kenya Colony*. Cambridge: *Cambridge University Press*.

Leakey, L. S. B. 1936. *Stone Age Africa*. London: *Oxford University Press*.

Leakey, L. S. B. 1959. "A New Fossil Skull From Olduvai." *Nature* 184: 491-493.

Leakey, L. S. B. and M. D. Leakey. 1964. "Recent Discoveries of Fossil Hominids in Tanganyika: At Olduvai and Near Lake Natron." *Nature* 202: 5-7.

Leakey, L. S. B., P. V. Tobias, and J. R. Napier. 1964. "A New Species of The Genus *Homo* From Olduvai Gorge." *Nature* 202: 7-9.

Leakey, Meave G., Fred Spoor, Frank H. Brown, Patrick N. Gathogo, Christopher Kiarie, Louise N. Leakey, and Ian McDougall. 2001. "New hominin genus from eastern Africa shows diverse middle Pliocene lineages." *Nature* 410: 433-440.

Leakey, Meave G., Craig S. Feibel, Ian McDougall, and Alan Walker. 1995. "New four-million-year-old hominid species from Kanapoi and Allia Bay, Kenya." *Nature* 376: 565-571.

Leakey, Meave G., Craig S. Feibel, Ian McDougall, C. Ward, and A. Walker. 1998. "New specimens and confirmation of an early age for Australopithecus anamensis." *Nature* 393, no. 6680: 62-6.

Leakey, R. E. F. 1973. "Evidence for an Advanced Plio-Pleistocene Hominid from East Rudolf, Kenya." *Nature* 242: 447-450.

Leakey, R. E. F. 1976. "New hominid fossils from the Koobi Fora formation in Northern Kenya." *Nature* 261: 574-576.

Lebatard, Anne-Elisabeth, Didier L. Bourlès, Philippe Duringer, M. Jolivet, R. Braucher, Julien Carcaillet, and Mathieu Schuster et al. 2008. "Cosmogenic nuclide dating of Australopithecus bahrelghazali and Sahelanthropus tchadensis: Plio-Miocene Hominids from Chad." *Proceedings of the National Academy of Sciences* 105, no. 9: 3226-31.

Leierer, Lucia, Ángel Carrancho Alonso, Leopoldo Pérez, Ángela Herrejón Lagunilla, Antonio V. Herrera-Herrera, Rory Connolly, and Margarita Jambrina-Enríquez et al. 2020. "It's getting hot in here - Microcontextual study of a potential pit hearth at the Middle Paleolithic site of El Salt Spain." *Journal of Archaeological Science* 123: 105237.

Lemke, Ashley. 2024. *Anthropological Theory Underwater.* Cambridge: Cambridge University Press, 2024: 32.

Lepre, Christopher J. and Dennis V. Kent. 2015. "Chronostratigraphy of KNM-ER 3733 and other Area 104 hominins from Koobi Fora." *Journal of Human Evolution* 86: 99-111.

Leroi-Gourhan, Arlette. 1975. "The Flowers Found with Shanidar IV, a Neanderthal Burial in Iraq." *Science* 190, no. 4214: 562-564.

Li, Zhan-Yang, Xiu-Jie Wu, Li-Ping Zhou, Wu Liu, Xing Gao, Xiao-Mei Nian, and Erik Trinkaus. 2017. "Late Pleistocene archaic human crania from Xuchang, China." *Science* 355, no. 6328: 969-972.

Linkov, Igor, S.E. Galaitsi, Benjamin D. Trump, Elizaveta Pinigina, Krista Rand, Eric H. Cline, and Maksim Kitsak. 2024. "Are civilizations destined to collapse? Lessons from the Mediterranean Bronze Age." *Global Environmental Change* 84, no. 102792.

Lloyd, Graeme T., Katie Elizabeth Davis, David Pisani, James E. Tarver, Marcello Ruta, Manabu Sakamoto, and David Hone et al. 2008. "Dinosaurs and the Cretaceous Terrestrial Revolution." *Proceedings of the Royal Society B* 275, no. 1650: 2483-90.

Lordkipanidze, David, Marcia S. Ponce de León, Ann Margvelashvili, Yoel Rak, G. Philip Rightmire, Abesalom Vekua, and Christoph P. E. Zollikofer. 2013. "A Complete Skull from Dmanisi, Georgia, and the Evolutionary Biology of Early *Homo*." *Science* 342, no. 6156: 326-331.

Loron, Corentin C., Laura M. Cooper, Sean McMahon, Seán F. Jordan, Andrei V. Gromov, Matthews Humpage, and Laetitia Pichevin et al. 2025. "Prototaxites was an extinct lineage of multicellular terrestrial eukaryotes." *bioRxiv* 643340.

Lyell, Charles. 1863. *The Geological Evidences of the Antiquity of Man: with remarks on theories on the origin of species by variation*. Ch. 21: 411. London: John Murray.

Macho, Gabriele A. 2015. "Pliocene hominin biogeography and ecology." *Journal of Human Evolution* 87: 78-86.

Maglio, V. J. 1972. "Vertebrate faunas and chronology of hominid-bearing sediments east of Lake Rudolf, Kenya." *Nature* 239, no. 5372: 379-85.

Martz, A. M. and F. H. Brown. 1981. "Chemistry and mineralogy of some Plio-Pliestocene tuffs from the Shungura Formation, southwest Ethiopia." *Quaternary Research* 16, no. 2: 240-257.

Mayr, Ernst. 1950. "Taxonomic Categories in Fossil Hominids." *Cold Springs Harbor Symposia Quant Biol* 15: 109-118.

McDougall, Ian. 1981. "40Ar/39Ar age spectra from the KBS Tuff, Koobi Fora Formation." *Nature* 294: 120-124.

McDougall, Ian. 1985. "K-Ar and 40Ar/39Ar dating of the hominid-bearing Pliocene-Pleistocene sequence at Koobi Fora, Lake Turkana, northern Kenya." *GSA Bulletin* 96, no. 2: 159-175.

McDougall, Ian. 2006. "Precise 40Ar/39Ar geochronology for the upper Koobi Fora Formation, Turkana Basin,

Northern Kenya." *Journal of the Geological Society* 163(1).

McHenry, Lindsay J. 2012. "A revised stratigraphic framework for Olduvai Gorge Bed I based on tuff geochemistry." *Journal of Human Evolution* 63, no. 2: 284-299.

McNutt, Ellison J., Kevin G. Hatala, Catherine Miller, James Adams, Jesse Casana, Andrew S. Deane, and Nathaniel J. Dominy et al. 2021. "Footprint evidence of early hominin locomotor diversity at Laetoli, Tanzania." *Nature* 600: 468-471.

McPherron, Shannon P., Zeresenay Alemseged, Curtis W. Marean, Jonathan G. Wynn, Denné Reed, Denis Geraads, and René Bobe et al. 2010. "Evidence for stone-tool-assisted consumption of animal tissues before 3.39 million years ago at Dikika, Ethiopia." *Nature* 466: 857-860.

Meredith, Robert W., Jan E. Janečka, John Gatesy, Oliver A. Ryder, Colleen A. Fisher, Emma C. Teeling, and Alisha Goodbla et al. 2011. "Impacts of the Cretaceous Terrestrial Revolution and KPg Extinction on Mammal Diversification." *Science* 334, no. 6055: 521-524.

Mojzsis, Stephen J., T. Mark Harrison, and Robert T. Pigeon. 2001. "Oxygen-isotope evidence from ancient zircons for liquid water at the Earth's surface 4,200 Myr ago." *Nature* 409: 178-181.

Momber, Garry. 2014. "Submerged Archaeology and Cultural Responses to Climatic Amelioration." In *Prehistoric Archaeology on the Continental Shelf*, edited by

Amanda M. Evans, Joseph C. Flatman, and Nicholas C. Flemming, 197, Springer, 2014.

Moore, M. W., T. Sutikna, Jatmiko, M. J. Morwood, and A. Brumm. 2009. "Continuities in stone flaking technology at Liang Bua, Flores, Indonesia." *Journal of Human Evolution* 57, no. 5: 503-526.

Mounier, Aurélien, François Marchal, and Silvana Condemi. 2009. "Is *Homo heidelbergensis* a distinct species? New insight on the Mauer mandible." *Journal of Human Evolution* 56, no. 3: 219-46.

Murphy, William J., Jan E. Janečka, Tanja Stadler, Eduardo Eizirik, Oliver A. Ryder, John Gatesy, and Robert W. Meredith et al. 2012. "Response to Comment on "Impacts of the Cretaceous Terrestrial Revolution and KPg Extinction on Mammal Diversification." *Science* 337, no. 6090: 34.

Muttoni, Giovanni and Dennis V. Kent. 2024. "Hominin population bottleneck coincided with migration from Africa during the Early Pleistocene ice age transition." *PNAS* 121, no. 13: e2318903121.

Nel, André, Jakub Prokop, Martina Pecharová, Micheal S. Engel, and Romain Garrouste. 2018. "Palaeozoic giant dragonflies were hawker predators." *Scientific Reports* 8, no. 12141.

Nerlich, Andreas G., Eduard Egarter Vigl, Angelika Fleckinger, Martina Tauber, and Oliver Peschel. 2021. "The Iceman: Life scenarios and pathological findings from 30 years of research on the glacier mummy "Ötzi." *Pathologe* 42, no. 5: 530-539.

Ni, Xijun, Qiang Ji, Wensheng Wu, Qingfeng Shao, Yannan Ji, Chi Zhang and Lei Liang et al. 2021. "Massive cranium from Harbin in northeastern China establishes a new Middle Pleistocene human lineage." *The Innovation* 2, no. 3: 100130.

O'Dea, Aaron, Harilaos A. Lessios, Anthony G. Coates, Ron I. Eytan, Sergio A. Restrepo-Moreno, Alberto L. Cione, and Laurel S. Collins et al. 2016. "Formation of the Isthmus of Panama." *Science Advances* 2, no. 8: e1600883.

Oktaviana, Adhi Agus, Renaud Joannes-Boyau, Budianto Hakim, Basran Burhan, Ratno Sardi, Shinatria Adhityatama, and Hamrullah et al. 2024. "Narrative cave art in Indonesia by 51,200 years ago." *Nature* 631: 814-818.

Oparin, A. I. 1938. *The Origin of Life* (trans: Sergius Morgulis). Macmillan, New York.

Oxford University Museum of Natural History. 2025. "Mary Anning's Ichthyosaur." *Ox.ac.uk.*

Pabst, M. A., I. Letofsky-Papst, E. Bock, M. Moser, L. Dorfer, E. Egarter-Vigl, and F. Hofer. 2009. "The tattoos of the Tyrolean Iceman: a light microscopical, ultrastructural and element analytical study." *Journal of Archaeological Science* 36, no. 10: 2335-2341.

Pickering, Robyn, Jan D. Kramers, Philip John Hancox, Darryl J. de Ruiter, and Jon D. Woodhead. 2011b. "Contemporary flowstone development links early hominin bearing cave deposits in South Africa." *Earth and Planetary Science Letters* 306, no. 1-2: 23-32.

Pickering, Robyn, Paul H. G. M. Dirks, Zubair Jinnah, Darryl J de Ruiter, Steven E. Churchil, Andy R. Herries, and Jon D. Woodhead et al. 2011a. "*Australopithecus sediba* at 1.977 Ma and implications for the origins of the genus *Homo.*" *Science* 333, no. 6048: 1421-3.

Pickering, Travis Rayne, Jason L. Heaton, Ronald J. Clarke, Morris B. Sutton, C. K. Brain, and Kathleen Kuman. 2012. *Journal of Human Evolution* 62, no. 5: 618-628.

Pickford, Martin, Brigitte Senut, Dominique Gommery, and Jacques Treil. 2002. "Bipedalism in Orrorin tugenensis revealed by its femora." *Comptes Rendus Palevol* 1, no. 4: 191-203.

Pickford, Martin and Brigitte Senut. 2001. "The geological and faunal context of Late Miocene hominid remains from Lukeino, Kenya." *Comptes Rendus de l'Académie des Sciences - Series IIA - Earth and Planetary Science* 332, no. 2: 145-152.

Pigati, Jeffrey S., Kathleen B. Springer, Jeffrey S. Honke, David Wahl, Marie R. Champagne, Susan R. H. Zimmerman, and Harrison J. Gray et al. 2023. "Independent age estimates resolve the controversy of ancient human footprints at White Sands." *Science* 382, no. 6666: 73-75.

Playfair, John. 1802. *Illustrations of the Huttonian Theory of the Earth.* London: Cadell and Davies; Edinburgh: William Creech.

Plummer, Thomas W., James S. Oliver, Emma M. Finestone, Peter W. Ditchfield, Laura C. Bishop, Scott A. Blumenthal, and Cristina Lemorini et al. 2023. "Expanded geographic distribution and dietary strategies of the earliest Oldowan hominins and *Paranthropus*." *Science* 379, no. 6632: 561-566.

Pomeroy, Emma, Paul Bennett, Chris O. Hunt, Tim Reynolds, Lucy Farr, Marine Frouin, and James Holman et al. 2020. "New Neanderthal remains associated with the 'flower burial' at Shanidar Cave." *Antiquity* 94, no. 373: 11-26.

Pomeroy, Emma, Marta Mirazón Lahr, Federica Crivellaro, Lucy Farr, Tim Reynolds, Chris O. Hunt, and Graeme Barker. 2017. "Newly discovered Neanderthal remains from Shanidar Cave, Iraqi Kurdistan, and their attribution to Shanidar 3." *Journal of Human Evolution* 111: 102-118.

Pompeani, David P., Byron A. Steinman, Mark B. Abbott, Katherine M. Pompeani, William Reardon, Seth DePasqual, and Robin H. Mueller. 2021. "On the Timing of the Old Copper Complex in North America: A Comparison of Radiocarbon Dates from Different Archaeological Contexts." *Cambridge University Press: Radiocarbon* 63, no. 2: 513-531.

Pound, Matthew J. and Ulrich Salzmann. 2017. "Heterogeneity in global vegetation and terrestrial climate change during the late Eocene to early Oligocene transition." *Scientific Reports* 7, no. 43386.

Prinoth-Fornwagner, R., and Th.R. Niklaus. 1994. "The man in the ice: results from radiocarbon dating." *Nuclear Instruments and Methods in Physics Research Section B:*

Beam Interactions with Materials and Atoms 92, no. 1-4: 282-290.

Pšenička, Josef, Jiří Bek, Jiří Frýda, Viktor Žárský, Monika Uhlířová, and Petr Štorch. 2021. "Dynamics of Silurian Plants as a Response to Climate Changes." *Life (Basel)* 11, no. 9: 906.

Rafferty, John, Amy Tikkanen, and Gloria Lotha. 2013. "Nicolaus Steno." In *Encyclopedia Britannica*.

Rasmussen, Morten, Sara L. Anzick, Michael R. Waters, Pontus Skoglund, Michael DeGiorgio, Thomas W. Stafford Jr., and Simon Rasmussen et al. 2014. "The genome of a Late Pleistocene human from a Clovis burial site in western Montana." *Nature* 506: 225-229.

Reynolds, Tim, W. Boismeir, L. Farr, Chris Hunt, D. Abdulmatulb, and Graeme Barker. 2015. "New investigations at Shanidar Cave, Iraqi Kurdistan." *Antiquity: A Review of World Archaeology* 348.

Richmond, Brian G. and William Jungers. 2008. "Orrorin tugenensis Femoral Morphology and the Evolution of Hominin Bipedalism." *Science* 319: 5870.

Rightmire, Philip G. 1996. "The human cranium from Bodo, Ethiopia: evidence for speciation in the Middle Pleistocene." *Journal of Human Evolution* 31, no. 1: 21-39.

Roche, Damien, Loïc Ségalen, Brigitte Senut, and Martin Pickford. 2013. "Stable isotope analyses of tooth enamel carbonate of large herbivores from the Tugen Hills deposits: Palaeoenvironmental context of the earliest Kenyan hominids." *Earth and Planetary Science Letters* 381: 39-51.

Roksandic, Mirjana, Predrag Radović, and Joshua Lindal. 2018. "Revising the hypodigm of *Homo heidelbergensis*: A view from the Eastern Mediterranean." *Quaternary International* 466, Part A: 66-81.

Roksandic, Mirjana, Predrag Radović, Xiu-Jie Wu, and Christopher J. Bae. 2022. "*Homo bodoensis* and why it matters." *Evolutionary Anthropology* 31, no. 5: 240-244.

Roksandic, Mirjana, Predrag Radović, Xiu-Jie Wu, and Christopher J. Bae. 2021. "Resolving the "muddle in the middle": The case for *Homo bodoensis sp. Nov.*" *Evolutionary Anthropology* 31, no. 1: 20-29.

Rollo, Franco, Massimo Ubaldi, Luca Ermini, and Isolina Marota. 2002. "Ötzi's last meals: DNA analysis of the intestinal content of the Neolithic glacier mummy from the Alps." *PNAS* 99, no. 20: 12594-12599.

Roosevelt, Anna C., M. Lima Da Costa, C. Lopes Machado, M. Michab, N. Mercier, H. Valladas, and J. Feathers et al. 1996. "Paleoindian Cave Dwellers in the Amazon: The Peopling of the Americas." *Science* 272, no. 5260: 373-384.

Rose, Thomas, Stefano Natali, Andrea Brotzu, Shay Bar, and Yuval Goren. 2020. "First Evidence for alloying practices in the Chalcolithic Southern Levant (4500-3800 BCE) as revealed by metallography." *Heritage Science* 11, no. 193 (2023).

Rosell, Jordi and Ruth Blasco. 2019. "The early use of fire among Neanderthals from a zooarchaeological perspective." *Quaternary Science Reviews* 217: 268-283.

Rosenberg, Danny, Eli Buchman, Sariel Shalev, and Shay Bar. 2020. "A Large Copper Artefacts Assemblage of Fazael, Jordan Valley: new evidence of Late Chalcolithic Copper Metallurgy in the Southern Levant." *Documenta Praehistorica* 47 (2020): 246-261.

Sassaman, Kenneth E. 1993. "Distribution, Timing, and Technology of Early Pottery in the Southeastern United States." *Revista de Arqueología Americana* 14: 101-133.

Sawada, Yoshihiro, Martin Pickford, Brigitte Senut, Tetsumaru Itaya, Masayuki Hyodo, Tamaki Miura, and Chikako Kashine et al. 2002. "The age of Orrorin tugenensis, an early hominid from the Tugen Hills, Kenya." *Comptes Rendus Palevol* 1, no. 5: 293-303.

Schmölcke, U., E. Endtmann, S. Klooss, M. Meyer, D. Michaelis, B-H. Rickert, and D. Rößler. 2006. "Changes of sea level, landscape and culture: A review of the south-western Baltic Area between 8800 and 4000 BC." *Palaeogeography, Palaeoclimatology, and Palaeoecology* 240: 423-438.

Schoepfer, Shane D., Thomas J. Algeo, Bas van de Schootbrugge, and Jessica H. Whiteside. 2022. "The Triassic-Jurassic transition - A review of environmental change at the dawn of modern life." *Earth-Science Reviews* 232: 104099.

Schoetensack, Otto. 1908. "The lower jaw of *Homo heidelbergensis* from the sands of Mauer near Heidelberg. A contribution to the paleontology of man." Leipzig: *Engelmann*.

Scotese, Christopher R., Christian Vérard, Landon Burgener, Reece P. Elling, and Adam T. Kocsis. 2024. "The Cretaceous world: plate tectonics, palaeogeography and palaeoclimate." *Geological Society* 544: 31-202. London, Special Publications.

Semaw, Sileshi, P. Renne, J. W. K. Harris, C. S. Feibel, R. L. Bernor, N. Fesseha, and K. Mowbray. 1997. "2.5 million-year-old stone tools from Gona, Ethiopia." *Nature* 385: 333-336.

Semaw, Sileshi. 2000. "The World's Oldest Stone Artefacts from Gona, Ethiopia: Their Implications for Understanding Stone Technology and Patterns of Human Evolution Between 2.6-1.5 Million Years Ago." *Journal of Archaeological Science* 27, no. 12: 1197-1214.

Semaw, Sileshi, Michael J. Rogers, Jay Quade, Paul R. Renne, Robert F. Butler, Manuel Dominguez-Rodrigo, and Dietrich Stout et al. 2003. "2.6 million-year-old stone tools and associated bones from the OGS-6 and OGS-7, Gona, Afar, Ethiopia." *Journal of Human Evolution* 45, no. 2: 169-177.

Senut, Brigitte, Martin Pickford, Dominique Gommery, Pierre Mein, Kiptalam Cheboi, and Yves Coppens. 2001. "First hominid from the Miocene (Lukeino Formation, Kenya)." *Comptes Rendus de l'Académie des Sciences - Series IIA - Earth and Planetary Science* 332, no. 2: 137-144.

Shalev, S. and J.P. Northover. 1993. "The Metallurgy of the Nahal Mishmar Hoard Reconsidered." *Archaeometry* 35, no. 1: 35-47.

Sharma, Meenu. 2020. "The Indus Valley Civilization Urban Planning and Social Structures." *International Journal in Management and Social Science* 8, no. 3.

Shunkov, Michael V., Alexander Yu. Federochenko, Maxim B. Kozlikin, and Anatoly P. Derevianko. 2020b. "Initial Upper Paleolithic ornaments and formal bone tools from the East Chamber of Denisova Cave in the Russian Altai." *Quaternary International* 559: 47-67.

Shunkov, Michael V., Maxim B. Kozlikin, and Anatoly P. Derevianko. 2020a. "Dynamics of the Altai Paleolithic industries in the archaeological record of Denisova Cave." *Quaternary International* 559: 34-46.

Skinner, Matthew M., Nicholas B. Stephens, Zewdi J. Tsegai, Alexandra C. Foote, H. Huynh Nguyen, Thomas Gross, and Dieter Pahr et al. 2015a. "Human evolution. Human-like hand use in *Australopithecus africanus*." *Science* 347, no. 6220: 395-399.

Skinner, Matthew M. Nicholas B. Stephens, Zewdi J. Tsegai, Alexandra C. Foote, N. Huynh Nguyen, Thomas Gross, and Dieter Pahr et al. 2015b. "Response to Comment on "Human-like hand use in *Australopithecus africanus*." *Science* 348, no. 6239: 1101.

Smart, Matthew S., Gabriel Filippelli, William P. Gilhooly III, Kazumi Ozaki, Christopher T. Reinhard, John E. A. Marshall, and Jessica H. Whiteside. 2023. "The expansion of land plants during the Late Devonian contributed to the marine mass extinction." *Communications Earth & Environment* 4, no. 449.

Smith, Thierry, Kenneth D. Rose, and Philip D. Gingerich. 2006. "Rapid Asia-Europe-North America geographic dispersal of earliest Eocene primate *Teilhardina* during the Paleocene-Eocene Thermal Maximum." *PNAS* 103, no. 30: 11223-11227.

Smith, William. 1816. *Strata Identified by Organized Fossils*. Printed by W. Arding.

Soil Science Society of America. 2022. "Designations for Soil Horizons and Layers." In *Glossary of Soil Science Terms*, Appendix II: 146-149. SSSA.

Solecki, Ralph. 1971. "Shanidar: The First Flower People." New York: *Knopf*.

Spikins, Penny, Andy Needham, Barry Wright, Calvin Dytham, Maurizio Gatta, and Gail Hitchens. 2019. "*Living to fight another day*: The ecological and evolutionary significance of Neanderthal healthcare." *Quaternary Science Reviews* 217: 98-118.

Spikins, Penny, Andy Needham, Lorna Tilley, and Gail Hitchens. 2018. "Calculated or caring? Neanderthal healthcare in social context." *World Archaeology* 50, no. 3: 384-403.

Spoor, Fred, Meave G. Leakey, and Paul O'Higgins. 2016. "Middle Pliocene hominin diversity: *Australopithecus deyiremeda* and *Keyanthropus platyops*." *Philosophical Transactions of the Royal Society B: Biological Sciences* 371: 20150231.

Spoor, Fred, Meave G. Leakey, and Louise N. Leakey. 2010. "Hominin diversity in the Middle Pliocene of eastern Africa: the maxilla of KNM-WT 40000." *Philosophical*

Transactions of the Royal Society B: Biological Sciences 365, no. 1556: 3377-88

Stafford, Thomas W., P. E. Hare, Lloyd Currie, A. J. T. Jull, and Douglas J. Donahue. 1991. "Accelerator radiocarbon dating at the molecular level." *Journal of Archaeological Science* 18, no. 1: 35-72.

Steno, Nicolaus (Niels Steensen). 1667. *Elementorum Myologiae Specimen*. Florence: Joseph Cocchini.

Steno, Nicolaus (Niels Steensen). 1669. *Prodromus Dissertationis de Solido Intra Solidum Naturaliter Contento*. Florence: Steno.

Stollhofen, Harald, Ian G. Stanistreet, Nicholas Toth, Kathy D. Schick, Agata Rodríguez-Cintas, Rosa M. Albert, and Paul Farrugia et al. 2021. "Olduvai's oldest Oldowan." *Journal of Human Evolution* 150: 102910.

Strait, David S. and Frederick E. Grine. 2004. "Inferring hominoid and early hominid phylogeny using craniodental characters: the role of fossil taxa." *Journal of Human Evolution* 47, no. 6: 399-452.

Su, Denise F. 2013. "The Earliest Hominins: Sahelanthropus, Orrorin, and Ardipithecus." *Nature Education Knowledge* 4, no. 4: 11.

Su, Denise F. and Yohannes Haile-Selassie. 2022. "Mosaic habitats at Woranso-Mille (Ethiopia) during the Pliocene and implications for *Australopithecus* paleoecology and taxonomic diversity." *Journal of Human Evolution* 163: 103076.

Sümer, Arev P., Hélène Rougier, Vanessa Villalba-Mouca, Yilei Huang, Leonardo N. M. Iasi, Elena Essel, and Alba

Bossoms et al. 2024. "Earliest modern human genomes constrain timing of Neanderthal admixture." *Nature* 638: 711-717.

Sun, Chengsheng. 2024. "Zhoukoudian: the birthplace of paleoanthropology and the stimulus for Paleolithic archaeology in China." in *Geology's Significant Sites and their Contribution to Geoheritage*. London, Special Publications: *Geological Society* 543: 343-356.

Sun, Wei-dong, Li-peng Zhang, Jia Guo, Cong-ying Li, Yu-Hang Jiang, Robert E. Zartman, and Zhao-feng Zhang. 2016. "Origin of the mysterious Yin-Shang bronzes in China indicated by lead isotopes." *Scientific Reports* 6, no. 23304.

Tang, Qing, Wentao Zheng, Shuhan Zhang, Junxuan Fan, Leigh Anne Riedman, Xudong Hou, and A. D. Muscente et al. 2024. "Quantifying the Global Biodiversity of Proterozoic Eukaryotes." *Science* 386, no. 6728.

Tierney, Jessica E., Jiang Zhu, Mingsong Li, Andy Ridgwell, Gregory J. Hakim, Christopher J. Poulsen, and Ross D. M. Whiteford et al. 2022. "Spatial patterns of climate change across the Paleocene-Eocene Thermal Maximum." *Proc. Natl. Acad. Sci.* 119, no. 42: e2205326119.

Tirard, Stéphane. 2017. "J. B. S. Haldane and the origin of life." *J Genet.* 96, no. 5: 735-739.

Trinkaus, E., C. B. Stringer, C. B. Ruff, R. J. Hennessy, M. B. Roberts, and S. A. Parfitt. 1999. "Diaphyseal cross-sectional geometry of the Boxgrove 1 Middle Pleistocene human tibia." *Journal of Human Evolution* 37, no. 1: 1-25.

Trinkaus, E., and M. R. Zimmerman. 1982. "Trauma among the Shanidar Neanderthals." *American Journal of Physical Anthropology* 57, no. 1: 61-76.

Tsutaya, Takumi, Rikai Sawafuji, Alberto J. Taurozzi, Zandra Fagernäs, Ioannis Patramanis, Gaudry Troché, and Meaghan Mackie et al. 2025. "A male Denisovan mandible from Pleistocene Taiwan." *Science* 388, no. 6743.

University of California Museum of Paleontology. 2025. "Nicholas Steno." *Berkeley.edu*.

Vaccari, E. 2006. "The 'classification' of mountains in eighteenth century Italy and the lithostratigraphic theory of Giovanni Arduino (1714-1795)." In *The Origins of Geology in Italy*, edited by G.B. Vai, W. Glen, and E. Caldwell. Geological Society of America.

Vespasiani, Davide M., Guy S. Jacobs, Laura E. Cook, Nicholas Brucato, Matthew Leavesley, Christopher Kinipi, and François-Xavier Ricaut et al. 2022. "Denisovan introgression has shaped the immune system of present-day Papuans." *PLoS Genet* 18, no. 12: e1010470.

Villmoare, Brian, William H. Kimbel, Chalachew Seyoum, Christopher J. Campisano, Erin N. DiMaggio, John Rowan, and David R. Braun et al. 2015. "Early *Homo* at 2.8 Ma from Ledi-Geraru, Afar, Ethiopia." *Science* 347, no. 6228: 1352-1355.

Villmoare, Brian, Lucas K. Delezene, Amy L. Rector, Erin N. DiMaggio, Christopher J. Campisano, David A. Feary, and Baro'o Mohammed Ali et al. 2025. "New discoveries of *Australopithecus* and *Homo* from Ledi-Geraru, Ethiopia." *Nature*

Wagner, Günther A., Lutz Christian Maul, Manfred Löscher, and H. Dieter Schreiber. 2011. "Mauer - the type site of *Homo heidelbergensis*: paleoenvironment and age." *Quaternary Science Reviews* 30, no. 11-12: 1464-1473.

Walker, A., R. E. Leakey, J. M. Harris, and F. H. Brown. 1986. "2.5-Myr *Australopithecus boisei* from west of Lake Turkana, Kenya." *Nature* 322: 517-522.

Walker, James, David Clinnick, and Mark White. 2021. "We Are Not Alone: William King and the Naming of the Neanderthals." *American Anthropologist* 123, no. 4: 805-818.

Wang, Hua, Zhanyang Li, Haowen Tong, and Thijs van Kolfschoten. 2022. "Hominin paleoenvironment in East Asia: The Middle Paleolithic Xuchang-Lingjing (China) mammalian evidence." *Quaternary International* 633: 118-133.

Ward, C. V., Meave G. Leakey, and A. Walker. 2001. "Morphology of *Australopithecus anamensis* from Kanapoi and Allia Bay, Kenya." *Journal of Human Evolution* 41, no. 4: 255-368.

Waters, Michael R. 1992. *Principles of Geoarchaeology: A North American Perspective*. University of Arizona Press: 45-53, 262-269.

Waters, Michael R. and Thomas W. Stafford Jr. 2007. "Redefining the Age of Clovis: Implications for the Peopling of the Americas." *Science* 315: 1122.

Waters, Michael R. Thomas W. Stafford Jr., and David L. Carlson. 2020. "The age of Clovis - 13,050 to 12,750 cal yr B.P." *Science Advances* 6, no. 43.

Weaver, Lucas N., Julia R. Kelson, Robert M. Holder, Nathan A. Niemi, and Catherine Badgley. 2024. "On the role of tectonics in stimulating the Cretaceous diversification of mammals." *Earth-Science Reviews* 248: 104630.

Weiner, S., Q. Xu, P. Goldberg, J. Liu, and O Bar-Yosef. 1998. "Evidence for the use of fire at Zhoukoudian, China." *Science* 281, no. 5374: 251-3.

Welker, Frido, Jazmín Ramos-Madrigal, Petra Gutenbrunner, Meaghan Mackie, Shivani Tiwary, Rosa Rakownikow Jersie-Christensen, and Cristina Chiva et al. 2020. "The dental proteome of *Homo antecessor*." *Nature* 580: 235-238.

White, Tim D., Berhane Asfaw, Yonas Beyene, Yohannes Haile-Selassie, C. Owen Lovejoy, Gen Suwa, and Giday Woldegabriel. 2009. "*Ardipithecus ramidus* and the paleobiology of early hominids." *Science* 326, no. 5949: 75-86.

White, Tim D. and Gen Suwa. 1987. Hominid footprints at Laetoli: facts and interpretations." *American Journal of Physical Anthropology* 72, no. 4: 485-514.

White, Tim D., Gen Suwa, and Berhane Asfaw. 1994. "*Australopithecus ramidus*, a new species of early hominid from Aramis, Ethiopia." *Nature* 371: 306-312.

White, Tim D. and J. M. Harris. 1977. "Suid Evolution and Correlation of African Hominid Localities." *Science* 198, no. 4312: 13-21.

White, Tim D., Stanley H. Ambrose, Gen Suwa, and Giday Woldegabriel. 2010. "Response to Comment on the Paleoenvironment of Ardipithecus ramidus." *Science* 328, no. 5982: 1105-1105.

Wignall, Paul B. and Jed W. Atkinson. 2020. "A two-phase end-Triassic mass extinction." *Earth-Science Reviews* 208: 103282.

Wilde, Simon A., John W. Valley, William H. Peck, and Colin M. Graham. 2001. "Evidence from detrital zircons for the existence of continental crust and oceans on the Earth 4.4 Gyr ago." *Nature* 409: 175-178.

Williams, Scott A., Thomas Cody Prang, Marc R. Meyer, Thierra K. Nalley, Renier Van Der Merwe, Christopher Yelverton, and Daniel García-Martínez et al. 2021. "New fossils of *Australopithecus sediba* reveal a nearly complete lower back." *eLife* 10: e70447.

Wilson Mantilla, Gregory P., Stephen G. B. Chester, William A. Clemens, Jason R. Moore, Courtney J. Sprain, Brody T. Hovatter, and William S. Mitchell et al. 2021. "Earliest Palaeocene purgatoriids and the initial radiation of stem primates." *Royal Society Open Science* 8, no. 2: 210050.

Woldegabriel, Giday, Stanley H. Ambrose, Doris Barboni, Bonnefille Raymonde, Laurent Bremond, Brian S. Currie, and David DeGusta et al. 2009. "The Geological, Isotopic, Botanical, Invertebrate, and Lower Vertebrate Surroundings of Ardipithecus ramidus." *Science* 326, no. 5949: 65-65e5.

Wood, Bernard A. 2014. "Human evolution: Fifty years after *Homo habilis*." *Nature* 508: 31-33.

Wood, Bernard A. and David B. Patterson. 2020. "*Paranthropus* through the looking glass." *PNAS* 117, no. 38: 23202-23204.

Wu, Xiaohong, Chi Zhang, Paul Goldberg, David Cohen, Yan Pan, Trina Arpin, and Ofer Bar-Yosef. 2012. "Early pottery at 20,000 Years Ago in Xianrendong Cave, China." *Science* 336, no. 6089: 1696-1700.

Xia, Huan, Dongju Zhang, Jian Wang, Zandra Fagernäs, Ting Li, Yuanxin Li, and Juanting Yao et al. 2024. "Middle and Late Pleistocene Denisovan subsistence at Baishiya Karst Cave." *Nature* 632: 108-113.

Yee, Gordon T., Jeannine E. Eddleton, and Chris E. Johnson. 2004. "Copper Metal from Malachite circa 4000 B.C.E." *Journal of Chemical Education* 81, no. 2.

Zanolli, Clément, Yousuke Kaifu, Lei Pan, Song Xing, Armand S. Mijares, Ottmar Kullmer, and Friedemann Schrenk et al. 2022. "Further analyses of the structural organization of *Homo luzonensis* teeth: Evolutionary implications." *Journal of Human Evolution* 163, 103124.

Zavala, Elena, Zenobia Jacobs, Benjamin Vernot, Michael V. Shunkov, Maxim B. Kozlikin, Anatoly P. Derevianko, and Elena Essel et al. 2021. "Pleistocene sediment DNA reveals hominin and faunal turnovers at Denisova Cave." *Nature* 595, no. 7867: 399-403.

Zeberg, Hugo, Mattias Jakobsson and Svante Pääbo. 2024. "The genetic changes that shaped Neanderthals, Denisovans, and modern humans." *Cell* 187, no. 5: 1047-1058.

Zhang, Dong Ju, Huan Xia, Fahu Chen, Bo Li, Viviane Slon, Ting Cheng, and Ruowei Yang et al. 2020. "Denisovan DNA in Late Pleistocene sediments from Baishiya Karts Cave on the Tibetan Plateau." *Science* 370, no. 6516: 584-587.

Zhao, Qingpo, Huanhuan Ma, and Christopher J. Bae. 2020. "New discoveries from the early Late Pleistocene Lingjing site (Xuchang)." *Quaternary International* 563: 87-95.

INDEX

ABOUT THE AUTHOR

Ethan Mellon is a maritime archaeologist and wilderness first responder specializing in prehistoric geoarchaeology and submerged subsurface data interpretation. He has over half a decade of professional archaeological experience working in prehistoric and pre-contact sites throughout the Southeastern United States and Latin America. As a wilderness first responder, he strives to integrate the practical application of emergency rescue skills into archaeological fieldwork conducted in areas where primary care is more than an hour away.

Driven by a strong interest in deep time, his research focuses on reconstructing submerged paleolandscapes, tracking changing prehistoric environments, and analyzing how ancient cultures adapted to challenging terrain. In addition to his regional fieldwork in the Southeastern United States, he has archaeological and wilderness experience in Iceland, Cuba, Ecuador, Ireland, parts of the Mediterranean, and Northern Alaska.

In 2020, Ethan was awarded membership to the Lambda Alpha National Anthropology Honor Society just before earning his bachelor's degree in Anthropology and Archaeology from the University of North Carolina at Charlotte. After pursuing further independent field research and material culture preservation for several years after graduating, he attended the Maritime Studies Program at East Carolina University.

www.ingramcontent.com/pod-product-compliance
Lightning Source LLC
LaVergne TN
LVHW100247210926
848480LV00003B/855
* 9 7 9 8 2 3 4 1 3 3 0 3 8 *